Elements of
METRIC SPACES

Elements of
METRIC SPACES

Elements of
METRIC SPACES

MURSALEEN

Anamaya Publishers
New Delhi

Prof. Mursaleen Mohammad
Department of Mathematics
Aligarh Muslim University
Aligarh-202002, India
Currently, Department of Mathematics
Faculty of Science, P.O. Box 80203
King Abdul Aziz University
Jeddah, Saudi Arabia

ANAMAYA PUBLISHERS
F-230, Lado Sarai, New Delhi - 110 030, India
e-mail: anamayapub@vsnl.net
        anamayapub@indiatimes.com

ISBN 81-88342-42-4

Published by Manish Sejwal for Anamaya Publishers,
F-230, Lado Sarai, New Delhi-110 030. Printed in India.

Preface

This book, intended for students of mathematics at undergraduate level or beyond, is primarily a text for a course at the advanced undergraduate level and will also serve as a reference for those who need to use such a course in their work. The subject 'metric space' is not new and is one of the core courses in pure mathematics having tremendous applications in various branches of mathematics, and mathematical sciences. Besides reviving the old concepts along with detailed examples, some topics like linear metric spaces, FK-spaces, paranormed spaces and measure of noncompactness are special features of this book. Some important old concepts like semimetric spaces, relative compactness, equicontinuity, lower (upper) semicontinuous functions, uniform boundedness principle and Arzela-Ascoli's theorem, which have been overlooked by many authors are incorporated to make this book self-contained.

Beginning with preliminaries, Chapter 2 studies metric and semimetric spaces and other topological concepts like openness and closedness. Chapter 3 discusses completeness along with two famous theorems, viz. Cantor's intersection theorem and Baire's category theorem, while Chapter 4 deals with the concept of continuity together with the concepts of semi-continuous functions and uniform boundedness principle. Chapter 5 is on compactness along with the idea of equicontinuity and Arzela-Ascoli's theorem and Chapter 6 deals with connectedness. Some further topics on metric spaces, e.g. fixed point theorem and its application, and a new concept of measure of noncompactness, are described in Chapter 7 and concludes with Chapter 8 which is devoted to study some new concepts like paranormed spaces, FK-spaces and linear metric spaces.

The list of references have been a source of help in writing the present book.

I am very grateful to my teacher Prof. Z.U. Ahmad for his kind encouragements and thankful to my student Mr. Osama H. Edely, who read the entire manuscript and suggested a number of improvements. I would also like to express my sincere thanks to my wife for being great source of help and encouragement to complete the book.

MURSALEEN

Contents

1

Preliminaries

This chapter provides some basic definitions and notations used throughout the book.

1.1 Notations and Abbreviations

$\mathbb{N}$:   set of all natural numbers.

$\mathbb{Z}$ set of all integers.

$\mathbb{Q}$:   set of all rational numbers.

$\mathbb{Q}^*$:   set of all irrational numbers.

$\mathbb{R}$:   set of all real numbers.

$\mathbb{R}^+$:   set of positive real numbers.

$\mathbb{C}$:   set of all complex numbers.

K: set of all scalars real or complex.

$\lim\limits_{n}$: $\lim\limits_{n \to \infty}$

$\sum\limits_{k}$: $\sum\limits_{k=1}^{\infty}$

$\sup\limits_{n}$: $\sup\limits_{n \geq 0}$ supremum

iff or $\Leftrightarrow$: for if and only if,

$x = (x_k)$ or $\{x_k\}$: sequence whose kth term is x_k.

w: space of all sequences real or complex

1.2 Sets

Presuming that the reader has a basic knowledge of elementary set theory, we give here only some basic notations.

Let us recall that by a *set*, we always mean a well-defined collection of distinct objects called the *elements* or *members* of the set. Sometimes *class* or *family* instead of set shall be used.

If a set has a finite number of elements, it is called a *finite set*. An *infinite set* has an infinite number of elements.

Let A and B be any two sets. Then A is a *subset* of B if every element of A is also an element of B and is written as $A \subset B$. Here B is called a *superset* of A. Further, if $A \subset B$ and $A \neq B$, then A is called a *proper subset* of B. A set which has no elements is called an empty set denoted by $\varnothing$.

A set whose elements are used as names is called an *index set*. Which may be finite or infinite and is denoted by $\wedge$. For example, consider a sequence $\{x_n\}$ $(n = 1, 2, ...)$, i.e. $\{x_n : n \in \mathbb{N}\}$. Hence, the index set here is the set $\mathbb{N}$.

If $A \subset B$, then $B - A$ is called the *complement* of A with respect to B. In case B is taken as a universal set, $B - A$ is written as A^c and is simply read as the complement of A.

1.3   Union, Intersection and de Morgan's Law

Let $\mathfrak{S}$ be a family of sets A. Then, we define

$$\cup \{A : A \in \mathfrak{S}\} = \{x : x \in A \text{ for at least one } A \in \mathfrak{S}\},$$

$$\cap \{A : A \in \mathfrak{S}\} = \{x : x \in A \text{ for all } A \in \mathfrak{S}\}.$$

Sometimes, we write $\cup A_\alpha$, $\cap A_\alpha$, where α runs through some index set, i.e. $\alpha \in \wedge$. If α runs through $\mathbb{N}$, we usually write

$$\cup \{A_n : n \in \mathbb{N}\} = \bigcup_{n=1}^{\infty} A_n$$

Similarly we can write for $\bigcap_{n=1}^{\infty} A_n$.

If A, B and C are any three nonempty sets. Then

$$A \cup B = B \cup A \text{ and } A \cap B = B \cap A \qquad \text{(commutative law)}$$

$$\left. \begin{array}{l} A \cup (B \cup C) = (A \cup B) \cup C \\ A \cap (B \cap C) = (A \cap B) \cap C \end{array} \right\} \qquad \text{(associative law)}$$

$$\left. \begin{array}{l} A \cap (B \cup C) = (A \cap B) \cup (A \cap C) \\ A \cup (B \cap C) = (A \cup B) \cap (A \cup C) \end{array} \right\} \qquad \text{(distributive law)}$$

$$\left. \begin{array}{l} (A \cup B)^C = A^C \cap B^C \\ (A \cap B)^C = A^C \cup B^C \end{array} \right\} \qquad \text{(de Morgan's Law)}$$

Any two sets A and B are said to be *disjoint* if $A \cap B = \varnothing$.

1.4   Cartesian Product

Let x, y be any objects. Then the *ordered pair* (x, y) is defined as the set $\{\{x\}, \{x, y\}\}$.

Let X, Y be given sets. Then

$$X \times Y = \{(x, y): x \in X \text{ and } y \in Y\}$$

is called the *Cartesian product* of X and Y.

An *ordered n-tuple* $(x_1, x_2, \ldots, x_n)$ is an n-tuple of objects. If X_1, $X_2, \ldots, X_n$ are sets, we define the Cartesian product $X_1 \times X_2 \times \ldots \times X_n$ as the set of all ordered n-tuples $(x_1, x_2, \ldots, x_n)$, where $x_1 \in X_1$, $x_2 \in X_2, \ldots, x_n \in X_n$.

Example 1. In a system with two input channels, e.g., the inputs to a stereoamplifier, denote the input on channel #1 by x_1 and channel #2 by x_2, then the system input is the ordered pair (x_1, x_2).

Example 2. (i) $\mathbb{R} \times \mathbb{R}$ or $\mathbb{R}^2$ is the set of all ordered pairs of real numbers.

(ii) $\{1, 4\} \times \{3, 7\} = \{(1, 3), (1, 7), (4, 3), (4, 7)\}$.

(iii) $\mathbb{R}^n$ is the set of all ordered n-tuples.

1.5 Relations

Given a set X, a *relation* on X can be defined as any subset of the Cartesian product $X \times X$. If R is a relation on X and the ordered pair (x, y) is in R, we say that "x is related to y under the relation R" and we write xRy. If the ordered pair (x, y) is not in R, we say that "x is not related to y under the relation R" and write $x\!\!\not\!Ry$.

Example 3. Let X be a set of three boys: Ramu (aged 17), Ahmad (aged 14) and John, (aged 19). Let R be the relation on X defined by $x\,Ry$ iff x is younger than y. The subset R of $X \times X$ will then be made up of the ordered pairs:

(Ahmad, Ramu),   (Ahmad, John),   (Ramu, John).

A relation R on a set X is said to be *reflexive* if for each $x \in X$, one has xRx; i.e., each x is related to itself.

Example 4. Let $X = \mathbb{R}$. The relation ρ on $\mathbb{R}$ defined by $x \, \rho \, y \Leftrightarrow x \leq y$ is *reflexive*.

A relation R on a set X is said to be *symmetric* if xRy implies that yRx. A relation R on a set X is said to be *anti-symmetric* if xRy and yRx imply $x = y$ for all x and y in X.

Example 5. The relation ρ on $\mathbb{R}$ defined by $x\rho y \Leftrightarrow |x| = |y|$ is symmetric (also reflexive).

A relation R on the set X is said to be *transitive* if xRy and yRz imply that xRz.

Example 6. The relation ρ on $\mathbb{R}$ defined by $x\rho y \Leftrightarrow x < y$ is transitive, but neither symmetric nor reflexive.

A relation R on a set X is said to be an equivalence relation if it is reflexive, symmetric and transitive.

Example 7. Define a relation $\sim$ by $x \sim y$ such that $x - y$ is divisible by 2. It is easy to cheek that $\sim$ is an equivalence relation on $\mathbb{Z}$.

1.6   Functions.

Let X and Y be two sets. Suppose that there is a rule f which assigns to each element in X exactly one element of Y. We say that f is a Y-*valued function* defined on X, or f is a function from X into Y and is denoted as $f: X \to Y$. The set X is called the *domain* of f and Y the codomain of f. If $x \in X$, the corresponding element $y \in Y$ is called the f-*image* of x and is denoted by $f(x)$, i.e. $y = f(x)$; and x is called the *pre-image* of y. The set of all f-images of the elements of X is called the *range* of f, denoted by $f(X)$ and it is subset of Y. By $\mathscr{D}(f)$ we denote the domain of f.

A function g such that $\mathscr{D}(f) \subset \mathscr{D}(g)$ and $g(x) = f(x)$ for each $x \in \mathscr{D}(f)$ is said to be an *extension* of f. Now, let f be a function with $\mathscr{D}(f) = X$ and let A be a subset of X. A function h such that $\mathscr{D}(h) = A$ and $h(x) = f(x)$ for each $x \in A$ is said to be the *restriction* of f to A and is denoted by $f|_A$.

Let $f: X \to Y$. Then f is called *one-to-one* or 1-1 or *injective* if $f(x_1) = f(x_2)$ implies $x_1 = x_2$, for every $x_1, x_2 \in X$. If the range of f is the whole of Y, then f is called *onto* or *surjective*. A mapping which is both injective and surjective is called *bijective* or a *one-to-one correspondence* between A and B.

Example 8. $f: \mathbb{R} \to \{x \in \mathbb{R}: x \geq 0\}$, defined by $f(x) = x^2$ is surjective but not injective. On the other hand if we define $f: \mathbb{R}^+ \to \mathbb{R}^+$ by $f(x) = x^2$, then it is bijective.

Let $f: X \to Y$ be bijective. Since f is surjective, if $y \in Y$ then there exists $x \in X$ such that $y = f(x)$. This x is unique, since f is injective. Hence there is an *inverse function* $g: Y \to X$ such that $g(f(x)) = x$, for all $x \in X$, and $f(g(y)) = y$ for all $y \in Y$. It is usual to write $g = f^{-1}$.

Example 9. $f: \mathbb{R} \to \mathbb{R}^+$, defined by $f(x) = e^x$, is bijective. The inverse $f^{-1}: \mathbb{R}^+ \to \mathbb{R}$ is denoted by log.

1.7   Bounded and Continuous Functions

Let A and B be subsets of $\mathbb{R}$. A function $f: A \to B$ is said to be *bounded* on A if there is a positive real number M such that $|f(x)| \leq M$ for every $x \in A$. Note that if f is a bounded on A, then it is bounded on every subset $C \subset A$.

Let f be a real valued function defined on $\mathbb{R}$. f is said to be *continuous* at $x = a$ if given $\varepsilon > 0$ there exists a $\delta > 0$ such that $|f(x) - f(a)| < \varepsilon$ whenever $|x - a| < \delta$. The function f is said to be continuous on a subset $A \subset \mathbb{R}$ iff it is continuous at each point of A.

Note that if f is continuous on the closed and bounded interval $[a, b]$, then f is bounded on $[a, b]$.

1.8 Equivalent Sets

Sets X and Y are called *equivalent* $(X \sim Y)$ iff there exists a bijective map $f : X \to Y$. Note that $\sim$ is an equivalence relation.

Example 10. (i) $\mathbb{N} \sim \mathbb{Z}$
(ii) The interval $(-1, 1)$ in $\mathbb{R}$ is equivalent to $\mathbb{R}$.

1.9 Countable Sets

A set is called *countable* if it is equivalent to $\mathbb{N}$ or to a subset of $\mathbb{N}$. Otherwise it is called *uncountable*. In case the set is equivalent to $\{1, 2, \ldots, n\}$ it is called finite, with n elements.

Example 11. (i) Sets $\mathbb{N}$, $\mathbb{Z}$ and $\mathbb{Q}$ are countable.
(ii) The set $\mathbb{R}$ is uncountable.

1.10 Partial Ordering

Let X be a set. Then a *partial order relation* on X or a *partial ordering* of X, denoted in general by $<$, is a relation which is: (i) reflexive, (ii) transitive, (iii) antisymmetric.

A partial ordering $<$ on a set X is said to be a *linear ordering* (or simple ordering) of X if for any two elements x and y of X we have either $x < y$ or $y < x$.

A *total order relation* $<$ is a partial order relation with the additional property that for any x, y, either $x < y$ or $y < x$.

A *partially ordered set* is just a pair $(X, <)$, consisting of a set X and a partial order $<$ on it.

Example 12. (i) Let A be any collection of sets. Then the set inclusion $\subset$ is a partial order on A. It is not in general a total order on A

(ii) The relation $\leq$ is the total order on $\mathbb{R}$.
(iii) $\leq$ linearly orders the set $\mathbb{R}$.

1.11 Well Ordering and Well-Ordering Principle

A strict linear ordering $<$ on a set X is called a *well ordering* for X or is said to well order X if every non-empty subset of X contains a first element.

By the *first element* in $E \subset X$, we mean an element $a \in E$ such that $a \prec x$ whenever $x \in E$ and $x \neq a$, where $\prec$ is a partial order on X.

Well-Ordering Principle. Every set X can be well ordered; that is, there is a relation $<$ that well orders X.

1.12 Axiom of Choice

Let $\wedge$ be an index set. For every $\alpha \in \wedge$, let X_α be some non-empty set. Then there is a function f defined on $\wedge$ such that $f(\alpha) \in X_\alpha$ for all $a \in \wedge$.

1.13 Completeness Axiom

Let $\leq$ be a partial ordering on a set X and let $A \subset X$. We shall say that a point $x \in X$ is an *upper bound* for A if we have $a \leq x$ for all $a \in A$. If the set X itself has an upper bound y, we shall call y a *maximal element* for X. A number z is called a *least upper bound* for A if it is an upper bound for A and if $z \leq x$ for each upper bound x of A.

Clearly, the least upper bound of a set A is unique if it exists. The following axiom for real numbers simply guarantees its existence for sets with an upper bound.

Completeness Axiom. Every non-empty set S of real numbers which has an upper bound has a least upper bound.

1.14 Zorn's Lemma

Let X be a partially ordered set with the property that every totally ordered subset has an upper bound. Then X contains a maximal element.

Note that Zorn's lemma is equivalent to: (i) the axiom of choice, and (ii) the well-ordering principle.

1.15 Sequences and Series

A *sequence* is a function whose domain is $\mathbb{N}$, i.e., $f \colon \mathbb{N} \to \mathbb{R}$ (or $\mathbb{C}$). A sequence is usually denoted by $\{x_n\}$ without explicit reference to f, i.e. with each $n \in \mathbb{N}$ we associate a number $f(n)$ conventionally denoted by x_n. Also, we write

$$x = \{x_n\} \text{ or } (x_n)$$

$$= (x_1, x_2, \ldots)$$

The value x_n is called the nth term of the sequence (x_n).

A *series* is a pair of sequences (a, s), where $a = (a_n) = (a_1, a_2, \ldots)$ is given, and $s = (s_n)$ is related to a by $s_n = a_1 + a_2 + \ldots + a_n$. Usually we write, as is conventional, $\sum_k a_k$ instead of (a, s) and speak of the series $\sum_k a_k$.

1.16 Monotonic Increasing and Decreasing Sequences

Let (x_n) be a sequence of real numbers. If $x_n \leq x_{n+1}$ for $n = 1, 2, \ldots$, then (x_n) is called a *monotonic increasing* sequence. If $x_n < x_{n+1}$ for $n = 1, 2, \ldots$, then (x_n) is called a *strictly increasing* sequence.

Similarly, $\geq$ and $>$ are used respectively to define *monotonic decreasing* and *strictly decreasing* sequences.

A *monotonic* sequence is a sequence which is either monotonic increasing or monotonic decreasing.

If (x_n) is a given sequence and (n_k) is an increasing (strictly) sequence of positive integers, then (x_{n_k}) is called a subsequence of (x_n).

1.17 Bounded, Convergent and Cauchy Sequences

A sequence $x = (x_k)$ is called *bounded* if there exists $M \geq 0$ such that $|x_k| \leq M$ for all $k = 1, 2, \ldots$. We denote the set of all bounded sequences by l_∞.

x is said to be *convergent* (to the limit l) if for every $\varepsilon > 0$ there exists $N = N(\varepsilon)$ such that $|x_k - l| < \varepsilon$, for all $k \geq N$. We write $x_k \to l$ $(k \to \infty)$, or $\lim_k x_k = l$. We denote the set of all convergent sequences by c; and the set of all null sequences by c_0, i.e., $l = 0$.

Note that a bounded monotonic sequence is convergent.

x is called a *Cauchy sequence* if for all $\varepsilon > 0$, there exists $N = N(\varepsilon)$ such that $|x_n - x_m| < \varepsilon$ for all $n, m > N$, i.e. $|x_n - x_m| \to 0$ as $m, n \to \infty$. We denote the set of all Cauchy sequences by $\mathscr{C}$.

Note that $c = \mathscr{C}$ due to the fact that $\mathbb{R}$ with the usual metric is complete (cf. Section 3.2) .

1.18 Convergent and Absolutely Convergent Series

A series $\sum_k x_k$ is said to be *convergent* if $s \in c$, where $s = (s_n)$, $s_n = x_1 + x_2 + \ldots + x_n$. The sum of the series $\sum_k x_k = \lim_n s_n$ when the series is convergent. We denote the set of all convergent series by cs.

A series $\sum_k x_k$ is said to be *absolutely convergent* if $\sum_k |x_k|$ converges. We denote the set of all absolutely convergent series by l_1.

Note that $l_1 \subset cs \subset c_0 \subset c \subset l_\infty$, all the inclusions being strict.

1.19 Sequence of Functions: Pointwise Convergence and Uniform Convergence

In place of sequences of real numbers, we consider the sequences of functions which are denoted by $\{f_n(x)\}$ or $\{f_n\}$.

A sequence of functions $\{f_n(x)\}$, with domain D, is said to be *convergent* at $x_0 (x_0 \in D)$ if the sequence $\{f_n(x_0)\}$ is convergent. The sequence $\{f_n(x)\}$ is said to be *pointwise convergent* on $S \subset D$, if $\{f_n(x_0)\}$ converges for each $x_0 \in S$.

A sequence $\{f_n(x)\}$ is said to be *uniformly convergent* to $f(x)$ on S if, given $\varepsilon > 0$ there exists N (depending upon ε only) such that $|f_n(x) - f(x)| < \varepsilon$ for all $n \geq N$ and for every $x \in S$. $f(x)$ is called the limit function.

Note that a sequence of continuous functions may have a limit function discontinuous.

1.20 Convergence Almost Everywhere

A sequence of functions $\{f_n\}$ is said to converge to f *almost everywhere*, i.e.

$$\lim_n f_n = f \,(\text{a.e.})$$

if the set of real numbers T, for which $\{f_n(t)\}$ fails to converge to $f(t)$, has Lebesgue measure zero.

In general, a property $\wp$ is said to hold almost everywhere on a set S, if the set of points of S, where $\wp$ does not hold, has Lebesgue measure zero.

1.21 Essential Supremum

Let f be a measurable function. Then

$$\inf\{\alpha > 0 : f \leq \alpha \text{ a.e.}\}$$

is called the *essential supremum* of f denoted by ess.sup f.

1.22 Lebesgue Outer Measure

Let $E \subset \mathbb{R}$, then

$$m^*(E) = \inf\left\{\sum_{n=1}^{\infty} l(I_n): E \subset \bigcup_{n=1}^{\infty} I_n\right\}$$

is called the *Lebesgue Outer Measure* of E, where I_n is an open interval in $\mathbb{R}$ and $l(I_n)$ is the length of I_n.

Remark 1. (i) When E is bounded, say $E \subset [a, b]$, then $0 \leq m^*(E) \leq b - a$.

(ii) If $m^*(E) = 0$, E is said to be a set of measure zero.

(iii) $A \subset B \Rightarrow m^*(A) \leq m^*(B)$. This property is called *monotonicity*.

(iv) Let $\{F_n\}$ be a sequence of sets in $\mathbb{R}$. Then

$$m^*\left(\bigcup_{n=1}^{\infty} F_n \right) \leq \sum_{n=1}^{\infty} m^*(F_n)$$

This is called *countable subadditivity*.

1.23 Lebesgue Measure

A set $E \subset \mathbb{R}$ is said to be (*Lebesgue*) *measurable* if

$$m^*(A) = m^*(A \cap E) + m^*(A \cap E^c)$$

for every $A \subset \mathbb{R}$, where E^c is the complement of E in $\mathbb{R}$.

If E is measurable, then $m^*(E)$ is called *Lebesgue measure* of E and will be denoted by $m(E)$.

Remark 2. (i) $\mathbb{R}$ is itself a measurable set and complements and countable unions of measurable sets are measurable.

(ii) Since open intervals are seen to be measurable, every open set, being a (disjoint) countable union of open intervals, is measurable.

(iii) If E_1, E_2, ... are pairwise disjoint measurable sets then

$$m\left(\bigcup_{n=1}^{\infty} E_n \right) = \sum_{n=1}^{\infty} m(E_n)$$

1.24 Measurable Functions

Let f be an extended real-valued function (i.e. a function whose values are in $\mathbb{R} \cup \{\pm\infty\}$) whose domain is measurable then the following statements are equivalent:

(3.1) For each real number α, the set $\{x : f(x) > \alpha\}$ is measurable.

(3.2) For each real number α, the set $\{x : f(x) \geq \alpha\}$ is measurable.

(3.3) For each real number α, the set $\{x : f(x) < \alpha\}$ is measurable.

(3.4) For each real number α, the set $\{x : f(x) \leq \alpha\}$ is measurable.

These statements imply that

(3.5) For each extended real number α, the set $\{x : f(x) = \alpha\}$ is measurable.

An extended real-valued function f defined on a measurable set E is said to be *Lebesgue measurable* or briefly *measurable* on E if it satisfies one of the above four statements.

Remark 3. A continuous function defined on a measurable set is measurable but converse need not be true.

1.25   Characteristic and Simple Functions

For $E \subset \mathbb{R}$, the function χ_E defined by

$$\chi_E = \begin{cases} 1, & \text{if } x \in E, \\ 0, & \text{if } x \notin E, \end{cases}$$

is called the *characteristic function* of E.

Remark 4. χ_E is measurable iff E is measurable. A linear combination

$$\phi(x) = \sum_{i=1}^{n} a_i \chi_{E_i}(x)$$

is called a *simple function* if the sets $E_i = \{t \in \mathbb{R}: \phi(t) = a_i\}$ are disjoint and measurable, where $a_1, a_2, \ldots, a_n$ are the distinct nonzero values of ϕ.

A simple function is a real-valued function on $\mathbb{R}$ whose range is finite.

1.26   Lebesgue Integral

If f is a bounded measurable function defined on a measurable set E with $m(E) < + \infty$, then the *Lebesgue integral* of f is defined by

$$\int_E f(x)dx = \inf \int_E \phi(x)dx$$

for all simple functions $\phi \geq f$.

Note that $\displaystyle\int_E f = \int_{\mathbb{R}} f \cdot \chi_E$

Remark 5. (i) If $m(E) = 0$ then $\displaystyle\int_E f = 0$.

(ii) If $f = 0$ (a.e.) on E then $\displaystyle\int_E f = 0$ and if $f = 1$ (a.e.) then

$$\int_E f = m(E)$$

1.27 $L^p(E)$-Spaces

Let E be a measurable set in $\mathbb{R}$ and f a measurable function on E. For $1 \leq p \leq \infty$, let

$$n_p(f) = \begin{cases} \left(\int |f|^p \, dm \right)^{1/p}, & \text{if } 1 \le p < \infty; \\ \text{ess.sup} |f|, & \text{if } p = \infty \end{cases}$$

If $n_p(f) < \infty$ for some p with $1 \le p < \infty$, then f is said to be *p-integrable*; 1-integrable and 2-integrable functions are known as *integrable* and *square integrable* functions, respectively. If $n_\infty(f) < \infty$ then f is said to be *essentially bounded*.

If we let $d_p(f, g) = n_p(f - g)$, $1 \le p \le \infty$, for measurable functions f, g on E, then d_p satisfies all the properties of a metric except that $d_p(f, g)$ may be zero without the function f being equal to g at all points of E. This difficulty is overcome by introducing the following equivalence relation on the set of all measurable functions on E. If f and g are measurable functions on E, then let $f \approx g$ if

$$m(\{t \in E: f(t) \ne g(t)\}) = 0$$

It is easily seen that if f is a nonnegative measurable function on E then $f \approx 0$ iff $\int_E f dm = 0$. If we identify two p-integrable (or essentially bounded) functions which belong to the same equivalence class, then d_p is indeed a metric on these spaces of functions. The set of all equivalence classes of p-integrable (or essentially bounded) functions is denoted by $L^p(E)$ (resp., $L^\infty(E)$).

By an abuse of notation we shall denote an element in $L^p(E)$ by f itself; such an f need only be defined a.e. on E.

Note that $L^p(E)$, $1 \le p \le \infty$, is a complete metric space.

1.28 Inequalities

1. **Triangle inequality.** Let $a, b \in \mathbf{K}$. Then $|a + b| \le |a| + |b|$.

2. Let $a, b \in \mathbf{K}$. Then $\dfrac{|a + b|}{1 + |a + b|} \le \dfrac{|a|}{1 + |a|} + \dfrac{|b|}{1 + |b|}$

3. Let $p > 1$, $p^{-1} + q^{-1} = 1$, and $a, b \ge 0$.

Then

$$ab \le \frac{a^p}{p} + \frac{b^q}{q}$$

with equality iff $a^p = b^q$.

4. Let $p \ge 1$. Then $\left(\sum_{k=1}^{n} |a_k| \right)^p \le n^{p-1} \sum_{k=1}^{n} |a_k|^p$.

5. Let $0 < p \le 1$, $a_1, a_2, \ldots, a_n \ge 0$ and $b_1, b_2, \ldots, b_n \ge 0$.

Then
$$\sum_{k=1}^{n} (a_k + b_k)^p \le \sum_{k=1}^{n} a_k^p + \sum_{k=1}^{n} b_k^p$$

6. Hölder Inequality. (i) Let $1 < p < \infty$ and $p^{-1} + q^{-1} = 1$.

For sums: Let $x = \{x_k\}$ and $y = \{y_k\}$ be real or complex sequences

such that
$$\sum_{k=1}^{\infty} |x_k|^p < \infty \quad \text{and} \quad \sum_{k=1}^{\infty} |y_k|^p < \infty$$

Then
$$\sum_{k=1}^{\infty} |x_k \, y_k| \le \left(\sum_{k=1}^{\infty} |x_k|^p \right)^{\frac{1}{p}} \left(\sum_{k=1}^{\infty} |y_k|^q \right)^{\frac{1}{q}}$$

For integrals: Let Ω be a bounded open set in $\mathbb{R}^n$, and f, g measurable
functions on Ω such that
$$\int_{\Omega} |f|^p \, dt < \infty \quad \text{and} \quad \int_{\Omega} |g|^q \, dt < \infty, t \in \Omega$$

Then
$$\int_{\Omega} |fg| \, dt \le \left(\int_{\Omega} |f|^p \, dt \right)^{\frac{1}{p}} \left(\int_{\Omega} |g|^q \, dt \right)^{\frac{1}{q}}$$

The special case $p = q = 2$ is often referred as the *Schwarz
inequality.*

(ii) Let $p = 1, q = \infty$

For sums: Given that
$$\sum_{k=1}^{\infty} |x_k| < \infty \quad \text{and} \quad \sup_k |y_k| < \infty$$

Then
$$\sum_{k=1}^{\infty} |x_k y_k| \le \left(\sum_{k=1}^{\infty} |x_k| \right) \left(\sup_k |y_k| \right)$$

For integrals: Given that
$$\int_{\Omega} |f| \, dt < \infty \quad \text{and} \quad \text{ess} \cdot \sup_t |g(t)| < \infty$$

Then
$$\int_{\Omega} |fg| \, dt \le \left(\int_{\Omega} |f| \, dt \right) \left(\text{ess} \cdot \sup_t |g(t)| \right)$$

7. Minkowski Inequality

For sums: Let $x = \{x_k\}$ and $y = \{y_k\}$ be real or complex sequences
such that, for $1 \le p < \infty$,
$$\sum_{k=1}^{\infty} |x_k|^p < \infty \quad \text{and} \quad \sum_{k=1}^{\infty} |y_k|^p < \infty$$

Then
$$\left(\sum_{k=1}^{\infty} |x_k \pm y_k|^p \right)^{\frac{1}{p}} \leq \left(\sum_{k=1}^{\infty} |x_k|^p \right)^{\frac{1}{p}} + \left(\sum_{k=1}^{\infty} |y_k|^p \right)^{\frac{1}{p}}$$

and for $p = \infty$

$$\sup_k |x_k + y_k| \leq \sup_k |x_k| + \sup_k |y_k|$$

if
$$\sup_k |x_k| < \infty \quad \text{and} \quad \sup_k |y_k| < \infty$$

For integrals: Let Ω be a bounded open set in $\mathbb{R}^n$, and f, g measurable functions on Ω such that

$$\int_\Omega |f|^p \, dt < \infty \quad \text{and} \quad \int_\Omega |g|^p \, dt < \infty$$

Then
$$\left(\int_\Omega |f \pm g|^p \, dt \right)^{\frac{1}{p}} \leq \left(\int_\Omega |f|^p \, dt \right)^{\frac{1}{p}} + \left(\int_\Omega |g|^p \, dt \right)^{\frac{1}{p}}$$

2

Metric and Semimetric Spaces

Metric space is a generalization of $\mathbb{R}$ (or $\mathbb{C}$), insofar as it is a space with a metric or a distance function. In $\mathbb{R}$ (or $\mathbb{C}$), the concept of absolute value $|x - x_0|$ plays an important role in defining two basic concepts—of convergence and continuity—on which the whole theory of real (or complex) variables depends. In the theory of metric spaces, the concept of distance is generalized by replacing $\mathbb{R}$ (or $\mathbb{C}$) with an arbitrary non-empty set X in such a way that we can have a notion of convergence and continuity in a more general setting.

2.1 Metric Spaces

Definition 1. Let X be a non-empty set. A metric on X is a function $d: X \times X \to \mathbb{R}$ satisfying the following conditions or axioms:

(M1) (positive): $d(x, y) \geq 0$ for all x and y in X,

(M2) (strictly positive): $d(x, y) = 0$ iff $x = y$ for all x and y in X,

(M3) (symmetry): $d(x, y) = d(y, x)$ for all x and y in X,

(M4) (triangle inequality): $d(x, y) \leq d(x, z) + d(z, y)$ for all x, y and z in X.

All the above four conditions are in harmony with our concept of distance in the Euclidean plane.

The set X with the metric d is called a *metric space* denoted by (X, d) and the set X is called the *underlying set*.

Definition 2. Let X be a non-empty set. Define, for $x, y \in X$

$$d(x, y) = \begin{cases} 0 & \text{if } x = y, \\ 1 & \text{if } x \neq y. \end{cases}$$

Then the metric d is called the *trivial metric* or *discrete metric* X_d on X.

A metric space (X, d) is, then a set X with an additional structure defined by means of a metric function d. This additional structure is called the *topological structure*. When there is no confusion we will often denote the metric space (X, d) by X.

Remarks: (i) A non-empty set X may have more than one metric defined on it. For example, if $X = \mathbb{R}^2$, the set of all ordered pairs of real numbers $x = (x_1, x_2)$,

$$d_1(x, y) = [(x_1 - y_1)^2 + (x_2 - y_2)^2]^{1/2}$$

or $$d_2(x, y) = |x_1 - y_1| + |x_2 - y_2|$$

are metrics on the set X. Hence (X, d_1) and (X, d_2) are different metric spaces even though they have the same underlying set X.

In general, any non-empty set with more than one element can have an infinite number of metrics defined on it. Indeed, if d is a metric on X, then $d_\alpha (x, y) = \alpha d(x, y)$, $\alpha > 0$, is a metric on X.

(ii) Condition (M1) follows from (M2), (M3) and (M4). For

$$d(x, y) + d(y, x) \geq d(x, x),$$

$\Rightarrow \quad 2d(x, y) \geq 0$, so that $d(x, y) \geq 0$ for all $x, y \in X$.

(iii) Let (X, d) be a metric space. Then

$$|d(x, y) - d(x', y')| \leq d(x, x') + d(y, y') \text{ for all } x, x', y, y' \in X.$$

It follows readily from the axioms.

2.2 Examples
Some examples are illustrations of the pure mathematical nature while others illustrate the occurrence of metric spaces in applied mathematics.

Example 1. The function d defined by $d(x, y) = |x - y|$ is a *natural* or *usual* metric for $\mathbb{R}$. The natural metric for $\mathbb{C}$ is $d(z, w) = |z - w|$, where $|z|$ is the usual modulus for $z \in \mathbb{C}$. Conditions (M1)-(M4) are easily verified.

Example 2. The set of n-tuples of complex numbers $\mathbb{C}^n$, is a metric space with

$$d(x, y) = \left(\sum_{k=1}^{n} |\xi_k - \eta_k|^2 \right)^{1/2}, \text{ where } x = (\xi_k), y = (\eta_k) \in \mathbb{C}^n$$

(M1)-(M3) can easily be verified. For (M4), we have to show that

$$\left(\sum_{k=1}^{n} |\xi_k - \eta_k|^2 \right)^{1/2} \leq \left(\sum_{k=1}^{n} |\xi_k - \lambda_k|^2 \right)^{1/2} + \left(\sum_{k=1}^{n} |\lambda_k - \eta_k|^2 \right)^{1/2} \quad (1)$$

Let us put $\xi_k - \lambda_k = \alpha_k$, $\lambda_k - \eta_k = \beta_k$ in (1). Then

$$\left(\sum_{k=1}^{n} |\alpha_k + \beta_k|^2 \right)^{1/2} \leq \left(\sum_{k=1}^{n} |\alpha_k|^2 \right)^{1/2} + \left(\sum_{k=1}^{n} |\beta_k|^2 \right)^{1/2}$$

Squaring both sides, we get

$$\left(\sum_{k=1}^{n} |\alpha_k + \beta_k|^2 \right) \leq \sum_{k=1}^{n} |\alpha_k|^2$$

$$+ 2 \left(\sum_{k=1}^{n} |\alpha_k|^2 \right)^{1/2} \left(\sum_{k=1}^{n} |\beta_k|^2 \right)^{1/2} + \sum_{k=1}^{n} |\beta_k|^2$$

that is

$$\sum_{k=1}^{n} |\alpha_k|^2 + 2\mathrm{Re}\left(\sum_{k=1}^{n} \alpha_k \overline{\beta}_k \right) + \sum_{k=1}^{n} |\beta_k|^2 \leq \sum_{k=1}^{n} |\alpha_k|^2$$

$$+ 2 \left(\sum_{k=1}^{n} |\alpha_k|^2 \right)^{1/2} \left(\sum_{k=1}^{n} |\beta_k|^2 \right)^{1/2} + \sum_{k=1}^{n} |\beta_k|^2$$

$$\Rightarrow \qquad \mathrm{Re}\left(\sum_{k=1}^{n} \alpha_k \overline{\beta}_k \right) \leq \left(\sum_{k=1}^{n} |\alpha_k|^2 \right)^{1/2} \left(\sum_{k=1}^{n} |\beta_k|^2 \right)^{1/2} \qquad (2)$$

Hence (1) is equivalent to (2).

Let us put $\theta_k = \dfrac{\alpha_k}{\left(\sum_{k=1}^{n} |\alpha_k|^2 \right)^{1/2}}$ and $\overline{\mu}_k = \dfrac{\beta_k}{\left(\sum_{k=1}^{n} |\beta_k|^2 \right)^{1/2}}$. Then (2) is

equivalent to

$$\mathrm{Re} \sum_{k=1}^{n} \theta_k \overline{\mu}_k \leq 1 \qquad (3)$$

i.e. to prove (M4) we can show (3). Now

$$\sum_{k=1}^{n} |\theta_k|^2 = 1 \text{ and } \sum_{k=1}^{n} |\overline{\mu}_k|^2 = 1, \text{ but for } k = 1, 2, 3, \dots, n$$

$$|\theta_k|^2 - 2\,\mathrm{Re}\; \theta_k \overline{\mu}_k + |\mu_k|^2 = |\theta_k - \mu_k|^2 \geq 0$$

Therefore, $2 - 2\,\mathrm{Re} \sum_{k=1}^{n} \theta_k \overline{\mu}_k \geq 0 \Rightarrow \mathrm{Re}\sum_{k=1}^{n} \theta_k \overline{\mu}_k \leq 1$.

Hence (M4) is proved and $\mathbb{C}^n$ is a metric space.

Example 3. Let w be the set of all sequences $x = (x_n)$ real or complex. We can define the metric d_w on w as

$$d_w(x, y) = \sum_{n=1}^{\infty} \frac{1}{2^n} \frac{|x_n - y_n|}{1 + |x_n - y_n|} \qquad (3.1)$$

$$x = (x_n), \; y = (y_n) \in w.$$

Since $\dfrac{|x_n - y_n|}{1 + |x_n - y_n|} \leq 1$ for each $n = 1, 2, 3, \ldots$, the series on the right-hand side of (3.1) is dominated by $\sum_n \dfrac{1}{2^n}$ which is convergent. Hence, $d_w(x, y)$ is well defined.

(M1) It follows readily, since the right-hand side of (3.1) is a series of positive terms.

(M2) Since $x = y \Leftrightarrow x_n = y_n$ for all n, so that $|x_n - y_n| = 0$ iff $x = y$. Hence $d_w(x, y) = 0$ iff $x = y$.

(M3) It follows easily from the fact that $|x_n - y_n| = |y_n - x_n|$.

(M4) For this we use the inequality

$$\frac{|a + b|}{1 + |a + b|} \leq \frac{|a|}{1 + |a|} + \frac{|b|}{1 + |b|}$$

Put $a = x_n - z_n$, and $b = z_n - y_n$, where $x = (x_n)$, $y = (y_n)$ and $z = (z_n) \in w$. We get

$$\frac{|x_n - y_n|}{1 + |x_n - y_n|} \leq \frac{|x_n - z_n|}{1 + |x_n - z_n|} + \frac{|z_n - y_n|}{1 + |z_n - y_n|}$$

Hence $d_w(x, y) \leq d_w(x, z) + d_w(z, y)$.

Example 4. Let $l_1 := \left\{ x \in w : \sum_{n=1}^{\infty} |x_n| < \infty \right\}$, the space of absolutely convergent series.

(l_1, d) is a metric space with the metric defined by

$$d(x, y) = \sum_{n=1}^{\infty} |x_n - y_n|, \; \text{for } x = (x_n), \; y = (y_n) \in l_1)$$

It is an easy verification.

Example 5. Let us define the following subsets of w:

$l_\infty := \{ x = (x_n) : \sup |x_n| < \infty \}$, the set of bounded sequences,

$c := \{ x = (x_n) : |x_n - l| \to 0 (n \to \infty), l \in \mathbb{C} \}$, the set of convergent sequences,

$c_0 := \{ x = (x_n) : |x_n| \to 0 (n \to \infty) \}$, the set of null sequences.

The above three sets are metric spaces equipped with the metric

$$d(x, y) = d_\infty(x, y) = \sup |x_n - y_n|,$$

where $x = (x_n)$, $y = (y_n)$ are in one of the underlying sets in consideration. The present metric on l_∞, c or c_0 is referred as the *usual metric* or *sup-metric*.

Conditions (M1)-(M3) are routine verifications. For (M4), note that

$$|x_n - z_n| \leq |x_n - y_n| + |y_n - z_n| \leq d(x, y) + d(y, z)$$

and hence sup $|x_n - z_n| \leq d(x, y) + d(y, z).$

Example 6. Let us define the set, for $p > 0$

$$l_p: = \{x = (x_n): \sum_{n=1}^{\infty} |x_n|^p < \infty\}$$

The function

$$d_p(x, y) = \left[\sum_{i=1}^{\infty} |x_i - y_i|^p \right]^{1/p}, 1 \leq p < \infty$$

defines a metric on l_p.

It follows from the Minkowski's inequality that the series defining $d_p(x, y)$ always converges for $x = (x_n)$, $y = (y_n) \in l_p$. Furthermore, the property (M4) is satisfied.

Remark. Such sequence spaces often occur in engineering and science. For example, the mathematical model for the set of all inputs to a sample-data control system is often taken to be (l_2, d_2). Another example is to consider a typical point in (l_2, d_2) as the sequence of coefficients associated with the modes of vibration of a mechanical system which has an infinite number of modes (e.g. a plucked string).

Example 7. Let $X(n)$ be the set of all ordered n-tuples of 'zeros' and 'ones', for example

$$X(3) = \{000, 001, 010, 011, 100, 101, 110, 111\}.$$

For $x, y \in X(n)$, let $d(x, y) =$ number of places where x and y have different entries. For example, in $X(3)$

$$d(110, 110) = 0, d(010, 110) = 1 \text{ and } d(101, 010) = 3$$

It is easy to verify that $(X(n), d)$ is a metric space.

Such metric space occurs in switching and automata theory.

Example 8. Consider the following system of differential equations

$$\begin{cases} \dfrac{dx_1}{dt} = a_{11}x_1 + a_{12}x_2 \\ \\ \dfrac{dx_2}{dt} = a_{21}x_1 + a_{22}x_2 \end{cases} \tag{1}$$

This system can be used as a mathematical model for the electrical network shown as follows:

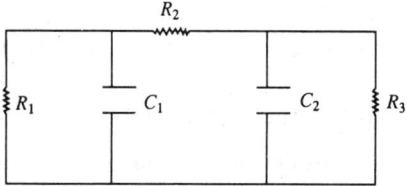

In this case,

$$a_{11} = -(R_1 + R_2)/C_1 \qquad a_{21} = R_2/C_2$$

$$a_{12} = R_2/C_1 \qquad a_{22} = -(R_2 + R_3)/C_2$$

Moreover, x_1 and x_2 are the voltages across the capacitors C_1 and C_2, respectively. Let us assume that $a_{11} = a_{22} = -2$ and $a_{12} = a_{21} = 1$, then the general solution of Eq. (1) is

$$\begin{cases} x_1(t) = \frac{1}{2}(e^{-t} + e^{-3t})x_1(0) + \frac{1}{2}(e^{-t} + e^{-3t})x_2(0) \\ x_2(t) = \frac{1}{2}(e^{-t} - e^{-3t})x_1(0) + \frac{1}{2}(e^{-t} + e^{-3t})x_2(0) \end{cases} \tag{2}$$

where $x_1(0)$ and $x_2(0)$ are the initial conditions.

Let X be the set of all ordered pairs (x_1, x_2) of bounded, continuous, real valued functions defined on the interval $0 \leq t < \infty$. Define

$$d(x, y) = \sup_t |x_1(t) - y_1(t)| + \sup_t |x_2(t) - y_2(t)|$$

Then it is easy to check that (X, d) is a metric space.

Now we consider some examples of function spaces.

Example 9. Let $X = C[0, T]$ be the set of all real-valued (or complex-valued) continuous functions on the interval $[0, T]$, $T > 0$. Define a metric on $C[0, T]$ by

$$d(f, g) = d_\infty(f, g) = \sup\{|f(t) - g(t)|: 0 \leq t \leq T\}$$

which is called the *sup-metric*.

Conditions (M1)-(M3) are easily verified. For (M4), let f, g and h be arbitrary elements of $C[0, T]$, then

$$d(x, y) = \sup_{0 \leq t \leq T} |f(t) - h(t) + h(t) - g(t)| \leq \sup_t |f(t) - h(t)|$$

$$+ \sup_t |h(t) - g(t)| = d(f, h) + d(h, g).$$

Hence $(C[0, T], d_\infty)$ is a metric space.

Example 10. Let $X = BC(I)$ be the set of all real or complex-valued, continuous, bounded functions f defined on the finite or infinite interval I. Define the metric on $BC(I)$ by

$$d_\infty(f, g) = \sup\{|f(t) - g(t)| : t \in I\}$$

called the *sup-metric*. Then $(BC(I), d_\infty)$ is a metric space.

Example 11. Let C be the unit circle in the complex plane, i.e. $C = \{z \in \mathbb{C} : |z| = 1\}$. Let X denote the set of all complex-valued functions $f(z)$ defined on C for which $\displaystyle\int_0 |f(z)|^2 |dz| < \infty$. Define

$$d(f, g) = \left(\int_c |f(z) - g(z)|^2 |dz| \right)^{1/2}$$

$$= \left(\int_0^{2\pi} |f(e^{i\theta}) - g(e^{i\theta})|^2 |d\theta| \right)^{1/2}$$

It is easy to check that (C, d) is a metric space.

Now we consider some *generalized sequence spaces*.

Definition 3. Let $p = (p_k)$ be a sequence of strictly positive numbers such that $0 < p_k \le \sup_k p_k = H < \infty$. Then we define the following sequence spaces

$$l(p) := \left\{ x = (x_k) : \sum_k |x_k|^{p_k} < \infty \right\}$$

$$l_\infty(p) := \left\{ x = (x_k) : \sup_k |x_k|^{p_k} < \infty \right\}$$

$$c_0(p) := \left\{ x = (x_k) : |x_k|^{p_k} \to 0 \text{ as } k \to \infty \right\}$$

$$c(p) := \left\{ x = (x_k) : |x_k - l|^{p_k} \to 0 \text{ as } k \to \infty \text{ for some } l \in \mathbb{C} \right\}$$

If (p_k) is constant and we write $p_k = p$ for all k, then the above spaces are reduced to l_p, l_∞, c_0 and c, respectively.

Theorem 1. The space $l(p)$ is a metric space with the metric defined by

$$d(x, y) = \left(\sum_k |x_k - y_k|^{p_k} \right)^{1/M},$$

$x = (x_k)$, $y = (y_k) \in l(p)$, where $M = \max(1, H)$.

Proof. Conditions (M1)-(M3) are easily verified. For (M4) note that $t_k = p_k/M \le 1$ and use the following inequality

$$|a_k + b_k|^{t_k} \le |a_k|^{t_k} + |b_k|^{t_k} \tag{1}$$

Since $M \ge 1$, the above inequality and the Minkowski's inequality imply that

$$\left(\sum_k |a_k + b_k|^{p_k} \right)^{1/M} \le \left(\sum_k |a_k|^{p_k} \right)^{1/M} + \left(\sum_k |b_k|^{p_k} \right)^{1/M} \tag{2}$$

Putting $a_k = x_k - y_k$ and $b_k = y_k - z_k$ in Eq. (2) we get

$$d(x, z) \le d(x, y) + d(y, z)$$

2.3 Semimetric Spaces

There are important examples where the function d defined on a non-empty set X is not exactly the metric but very near to it which motivates to give a slightly more general space than a metric space.

Definition 4. Let X be a non-empty set and $d: X \times X \to \mathbb{R}$ such that (M1), (M3) and (M4) of Definition 1 hold and (M2) is replaced by

$$(M2)' \quad d(x, x) = 0 \text{ for } x \in X$$

Then such function d is called *semimetric* or *pseudometric* on X and (X, d) a *semimetric space* or *pseudometric space*.

It is clear that every metric space is a semimetric space but not vice versa.

Example 12. Let $X = c$, be the space of convergent sequences. Define

$$d(x, y) = |\lim(x_n - y_n)| \text{ for } x = (x_n), y = (y_n) \in c$$

Then d is a semimetric but not metric. Since $d(x, x) = |\lim(x_n - x_n)|$ and if we take

$$x_n = 1/n, n = 1, 2, \dots$$

$$y_n = 0 \text{ for all } n$$

Then $d(x, y) = |\lim(1/n - 0)| = 0$, but $x \ne y$.

Example 13. Let $X = \mathbb{R}^2$. Define

$$d(x, y) = d\{(x_1, x_2), (y_1, y_2)\} = |x_1 - y_1|$$

for $x = (x_1, x_2), y = (y_1, y_2) \in \mathbb{R}^2$. Then we can easily check that d is a semi-metric on $\mathbb{R}^2$.

Example 14. Let $L = L[0, 1]$ be the set of Lebesgue integrable functions f on $[0, 1]$. Then $\int_0^1 |f(x)|dx < \infty$.

Define
$$d(f, g) = \int_0^1 |f(x) - g(x)|dx \quad (f, g \in L)$$

Then d is a semimetric on L and it fails to be a metric since $d(f, g) = 0$ implies only that $f = g$ almost everywhere on $[0, 1]$.

The following theorem provides a method to turn a semi-metric space into a metric space.

Theorem 2. Let (X, d) be a semimetric space and $\sim$ be an equivalence relation such that "$x \sim y$ to mean that $d(x, y) = 0$". If E_x is the equivalence class containing x, and $E = \{E_x : x \in X\}$, then

$$\rho(E_x, E_y) = d(x, y)$$

is well defined on $E \times E$ and (E, ρ) is a metric space.

Proof. The axioms for a semi-metric space imply that $\sim$ is an equivalence relation, e.g. $x\sim y$, $y\sim z$ imply $d(x, y) = d(y, z) = 0$ and since d is non-negative, we have

$$0 \le d(x, z) \le d(x, y) + d(y, z) = 0, \text{ i.e. } d(x, y) = 0$$

Hence $x \sim z$. To show that ρ is well-defined, let $x' \in E_x$, $y' \in E_y$. Then $x'\sim x$, $y'\sim y$, so that

$$|d(x, y) - d(x', y')| \le d(x, x') + d(y, y')$$

implies that $d(x, y) = d(x', y')$. Thus ρ is independent of which elements are chosen. Hence ρ is well-defined.

Now, if $E_x = E_y$, then $x \sim y$, $d(x, y) = 0$, $\rho(E_x, E_y) = 0$. Conversely, if $\rho(E_x, E_y) = 0$ then $d(x, y) = 0$, $x \sim y$ and so $E_x = E_y$. Hence (M2) holds. Here we used the fact that $E_x = E_y$ iff and only if $x\sim y$. So that we can easily verify (M3) and M(4).

Example 15. We can use Theorem 2 for Example 14. We have $f\sim g$ iff $d(f, g) = 0$, i.e. iff $f = g$ a.e. on $[0, 1]$. Thus L may be regarded as a metric space whose elements are equivalance classes of functions, where $g \in E_f$ means $g = f$ a.e. on $[0, 1]$.

2.4   Some Metric and Topological Concepts

The idea of neighbourhood plays a basic role in real and complex analysis. This section starts with the generalization of this notion for metric spaces which is subsequently used to define other metric and topological concepts.

Definition 5. Let (X, d) be a metric space and $a \in X$. Then for $r > 0$

$$B(a, r) \text{ or } S_r(a) := \{x \in X : d(a, x) < r\}$$

is called *open sphere* or *open ball* or *neighbourhood* with center a and radius r, and

$$B[a, r] \text{ or } S_r[a] := \{x \in X : d(a, x) \le r\}$$

is called *closed sphere* or *closed ball*.

Example 16. (i) In $\mathbb{R}$, the $S_r(a)$ and $S_r[a]$ are, respectively, the open interval $(a - r, a + r)$ and closed interval $[a - r, a + r]$.

(ii) In $\mathbb{C}$, the $S_r(z_0)$ and $S_r[z_0]$ are, respectively, the open disc $|z - z_0| < r$ and closed disc $|z - z_0| \le r$.

(iii) In $C[0, 1]$, the $S_1(\theta)$ is the set of all continuous functions lying (strictly) in a band of width 2 centered on the x-axis, where θ is the continuous function which is identically zero on $[0, 1]$.

(iv) Let a be any point in the discrete metric space X, then

$$S_r(a) := \begin{cases} \{a\}, & 0 < r \le 1 \\ X, & r > 1 \end{cases}$$

and

$$S_r[a] := \begin{cases} \{a\}, & 0 < r < 1 \\ X, & r \ge 1 \end{cases}$$

Definition 6. Let (X, d) be a metric space. Then $G \subset X$ is called *open* if each of its points is the center of some open sphere contained in G. Symbolically, for a given $x \in G$, there exists $r > 0$ such that $S_r(x) \subset G$.

Example 17. (a) $\mathbb{R}$ is an open set.

(b) $\mathbb{Q}$ is not an open set in $\mathbb{R}$.

(c) The set of irrationals is not open in $\mathbb{R}$.

(d) $[0, 1)$, $[0, 1]$ and $(0, 1]$ are not open in the usual metric space $\mathbb{R}$.

(e) Let $X = \{x : 0 \le x \le 1 \text{ or } 2 \le x \le 3\} = [0, 1] \cup [2, 3]$.

Let $d(x, y) = |x - y|$. Then the set $A = \{x : 0 \le x \le 1\}$ is an open subset of (X, d) but A is not an open subset of $(\mathbb{R}, d)$.

This shows that we have to be careful to state exactly which universe space X is being considered.

(f) $\{1/n : n \in \mathbb{N}\}$ is not an open set in $\mathbb{R}$.

(g) Every subset of a discrete metric space X_d is open. For, if $A \in X$, $a \in A$, then

$S_1(a) = \{a\} \subset A$, i.e. A is open.

(h) In a metric space (X, d), both $\emptyset$ and X are open.

Theorem 3. Let (X, d) be a metric space.

(a) Then each $S_r(a)$ is open in X.

(b) Let $G \subset X$. Then, G is open iff it is the union of open spheres.

Proof. (a) Let $x \in S_r(a)$ so that $d(x, a) < r$. Put $r' = r - d(x, a) > 0$. Let $y \in S_{r'}(x)$. Then $d(x, y) < r'$, and

$d(y, a) \le d(y, x) + d(x, a) < r' + d(x, a) = r$, i.e. $y \in S_r(a)$.

Hence $S_{r'}(x) \subset S_r(a)$. Thus for each point x in $S_r(a)$ there is an open sphere $S_{r'}(x)$ which is contained in $S_r(a)$, i.e. $S_r(a)$ is open.

(b) Let G be an open set. Then each point of G is the centre of some open sphere which is contained in G. Therefore, the set G is precisely the union of all such open spheres.

Conversely, let G be the union of open spheres and J be the family of these open spheres. Let $x \in G$ be arbitrary. Then, x must belong to some open sphere, say $S_r(x_0)$ of J.

Since every open sphere is an open set, x is the centre of an open sphere, say, $S_{r_1}(x)$ such that $S_{r_1}(x) \subset S_r(x_0)$.

Therefore $S_{r_1}(x) \subset J$, since $S_r(x_0) \subset J$. Hence G is open.

Theorem 4. Let (X, d) be a metric space. Then

(i) The union of any collection, i.e. arbitary union, of open sets is open.

(ii) The finite intersection of open sets is open but arbitrary intersection of open sets need not be open.

Proof. (i) Let $\wedge$ denote an index set and $\{G_\alpha\}_{\alpha \in \wedge}$ be a family of open sets in X. If $x \in \bigcup_{\alpha \in \wedge} G_\alpha$, then $x \in G_\alpha$ for some α, since each G_α is open. Thus there exists an open sphere $S_r(x) \subset G_\alpha$.

Hence $S_r(x) \subset G_\alpha \subset \bigcup_{\alpha \in \wedge} G_\alpha$, so $x \in G_\alpha$ implies that there exists $S_r(x) \subset \bigcup_{\alpha \in \wedge} G_\alpha$, so there exists an open sphere $S_r(x) \subset G$. Hence, $S_r(x) \subset G_\alpha \subset \bigcup_{\alpha \in \wedge} G_\alpha$, so $x \in \bigcup_{\alpha \in \wedge} G_\alpha$ implies tht there exists $S_r(x) \subset \bigcup_{\alpha \in \wedge} G_\alpha$ Therefore, $\bigcup_{\alpha \in \wedge} G_\alpha$ is open.

(ii) Let $\{G_i : i = 1, 2, ..., n\}$ be a finite collection of open sets in X and $x \in \bigcap_{i=1}^{n} G_i$ be arbitrary. Then $x \in G_i$ for each $i = 1, 2, ..., n$. Since each G_i is open, there exists an $r_i > 0$ such that $S_{r_i}(x) \subset G_i$, $i = 1, 2, ..., n$. Take $r = \min\{r_1, r_2, ..., r_n\}$. Then

$$S_r(x) \subset S_{r_i}(x) \subset G_i, i = 1, 2, ..., n$$

so that $S_r(x) \subset \bigcap\limits_{i=1}^{n} G_i$. Hence for $x \in \bigcap\limits_{i=1}^{n} G_i$ there is an open sphere

$S_r(x) \subset \bigcap\limits_{i=1}^{n} G_i$. Hence $\bigcap\limits_{i=1}^{n} G_i$ is open.

For arbitrary intersection, consider the family $\{(-1/n, 1/n): n \in \mathbb{N}\}$ of open sets in $\mathbb{R}$. Then we have $\cap \{(-1/n, 1/n) : n \in \mathbb{N}\} = \{0\}$ which is not open.

Definition 7. Let A be any subset of a metric space (X, d). A point $x \in A$ is called an *interior point* of A if there exists $r > 0$ such that $x \in S_r(x) \subseteq A$. A point $x \in A$ is said to be an *exterior point* of A, if it is an interior point of the complement of A.

The set A^0 of all interior points of A is called *interior* of A and the set ext(A) of all exterior points of A is called *exterior* of A. Note that

$$A^0 = \text{ext}(A^c) = (\overline{A^c})^c \quad \text{and} \quad (A^c)^0 = (\overline{A})^c$$

A point $x \in A$ is said to be a *frontier point* of $A \subseteq X$ if it is neither an interior nor an exterior point of A. If the interior point belongs to A it is then called a *boundary point* of A.

The set of all frontier points and boundary points are denoted by $F_r(A)$ and $bd(A)$, respectively. Clearly $bd(A) \subseteq F_r(A)$.

Example 18. (i) Let $X = \mathbb{R}$, and d the usual metric, $A = [0, 1)$, then

$$A^0 = (0, 1), \text{ext}(A) = (-\infty, 0) \cup (1, \infty),$$

$$F_r(A) = \{0, 1\} \text{ and } bd(A) = \{0\}$$

(ii) Let $A = \mathbb{Q}$ in (i), then

$$A^0 = \emptyset, \text{ext}(A) = \emptyset, F_r(A) = \mathbb{R} \text{ and } bd(A) = \emptyset.$$

(iii) If (X, d) is a discrete metric space and $A \subseteq X$, then

$$A^0 = A, \text{ext}(A) = A^c, F_r(A) = bd(A) = \emptyset$$

Theorem 5. Let (X, d) be a metric space. Then

(i) A^0 is open
(ii) A^0 is the largest open subset of A
(iii) A is open if and only if $A = A^0$
(iv) A^0 is the union of all open subsets of A
(v) $A \subseteq B \Rightarrow A^0 \subset B^0$
(vi) $(A \cap B)^0 = A^0 \cap B^0$
(vii) $A^0 \cup B^0 \not\subset (A \cup B)^0$

Proof. (i) Let $x \in A^0$. Then there exists an open sphere $S_r(x) \subset A$. Since $S_r(x)$ is open, each point of $S_r(x)$ is the interior point of A, i.e. $S_r(x) \subset A^0$. Hence A^0 is open.

(ii) Let $G \subset A$ be an open set and $x \in G$. Then there exists an open sphere $S_r(x) \subset G$. Therefore, $S_r(x) \subset A$ and by definition of A^0, we have $x \in A^0$. Hence, $G \subset A^0$. Thus for any given open set $G \subset A$, we obtain $G \subset A^0 \subset A$, where A^0 is open by part (i). Hence A^0 is the largest open subset of A.

(iii) and (iv) follow readily by (ii).

(v) Let $x \in A^0$. Then there exists $S_r(x) \subset A \subset B$, so that $x \in B^0$. Hence $A^0 \subset B^0$.

(vi) Let $x \in (A \cap B)^0$. Then there exists $S_r(x) \subset A \cap B$. Therefore $S_r(x) \subset A$ and $S_r(x) \subset B$, so that $x \in A^0$ and $x \in B^0$, Hence $x \in A^0 \cap B^0$, i.e. $(A \cap B)^0 \subset A^0 \cap B^0$.

Conversely, let $x \in A^0 \cap B^0$. Then $x \in A^0$ and $x \in B^0$, so there exist open spheres $S_{r_1}(x) \subset A$ and $S_{r_2}(x) \subset B$. If $r = \min \{r_1, r_2\}$, then $S_r(x) \subset A \cap B$. Therefore, $x \in (A \cap B)^0$, i.e. $A^0 \cap B^0 \subset (A \cap B)^0$. Combining two we get (vi).

(vii) $A^0 \cup B^0 \subset (A \cup B)^0$ can be proved easily. To prove that this inclusion is proper, let $X = \mathbb{R}$ and $A = [0, 1]$, $B = [1, 2]$. Then $A \cup B = [0, 2]$, $A^0 = (0, 1)$, $B^0 = (1, 2)$, and $(A \cup B)^0 = (0, 2)$. But $A^0 \cup B^0 = (0, 1) \cup (1, 2)$. Hence $A^0 \cup B^0 \subsetneq (A \cup B)^0$.

Definition 8. Let A be a subset of a metric space (X, d). Then a point $x \in X$ is called a *limit point* of A if each $S_r(x)$ contains at least one point of A other than x, i.e. if $(S_r(x) - \{x\}) \cap A \neq \emptyset$. The set of all limit points is called the *derived set*. A' denotes the derived set of A.

Sometimes the word accumulation point or cluster point is used in place of limit point.

Example 19. Let $X = \mathbb{R}$, the usual metric space and $A \subset X$. Then

(i) if $A = \mathbb{R}$, $\mathbb{Q}$ or the set of irrationals, then $A' = \mathbb{R}$.

(ii) if $A = \mathbb{Z}$, then $A' = \emptyset$

(iii) if $A = [a, b]$, $(a, b]$, $[a, b)$ or (a, b), then $A' = [a, b]$.

(iv) if $A = \{1, 1/2, 1/3, ...\}$, then $A' = \{0\}$.

(v) if $A = [0, 1] \cup \{2\}$, then $A' = [0, 1]$.

Definition 9. Let A be a subset of a metric space (X, d), then A is said to be *closed* if it contains all its limit points, i.e. if $A' \subset A$.

Sometimes following definition of closed set is suggested:
A in (X, d) is closed iff its complement in X is open.

Example 20. (i) Let X be a discrete metric space. Then every $A \subset X$ is closed. For A^c(the complement of A) $\subset X$. Hence, A^c is open by

Example 17(g). Thus, every subset of a discrete metric space is open and closed.

(ii) $\mathbb{Q}$ is not closed in $\mathbb{R}$, since $\mathbb{Q}' = \mathbb{R} \not\subset \mathbb{Q}$. The set of irrationals is also not closed.

(iii) $\mathbb{R}$ is closed, since $\mathbb{R}' = \mathbb{R}$.

(iv) If $A = \{1, 1/2, 1/3, ...\}$, then A is not closed in $\mathbb{R}$ since $A' = \{0\} \not\subset A$.

(v) $\emptyset$ and X both are closed in a metric space (X, d).

Remark. Note that "closed" is not to be interpreted as "not-open". There are examples where a set is both open and closed or neither open nor closed. In contrast of Theorem 4, we have the following:

Theorem 6. Let (X, d) be a metric space. Then
  (i) the arbitrary intersection of closed sets in X is closed.
  (ii) the finite union of closed sets in X is closed but the arbitrary union of closed sets need not be closed.

Proof. (i) Let $\{F_\alpha\}_{\alpha \in \wedge}$ be an arbitrary family of closed sets in X. Since each F_α is closed, so $G_\alpha = X - F_\alpha$ is open for each $\alpha \in \wedge$. Now, by DeMorgan's law

$$\bigcap_{\alpha \in \wedge} F_\alpha = \bigcap_{\alpha \in \wedge} (X - G_\alpha) = X - \bigcup_{\alpha \in \wedge} G_\alpha$$

By Theorem 4, $\bigcup_{\alpha \in \wedge} G_\alpha$ is open so that $X - \bigcup_{\alpha \in \wedge} G_\alpha$ is closed. Hence $\bigcap_{\alpha \in \wedge} F_\alpha$ is closed.

  (ii) Let $\{F_i : i = 1, 2, ...n\}$ be a family of closed sets. Then each $X - F_i = G_i$, is open so that

$$\bigcup_{i=1}^{n} F_i = \bigcup_{i=1}^{n} (X - G_i) = X - \bigcap_{i=1}^{n} G_i, \text{ by DeMorgan's law}$$

Again by Theorem 4, we have that $\bigcup_{i=1}^{n} F_i$ is closed.

For arbitrary union, consider the family $\{[1/n, 2]: n \in \mathbb{N}\}$ of closed sets in $\mathbb{R}$. Then

$$\cup\{[1/n, 2] : n \in \mathbb{N}\} = (0, 2] \text{ which is not closed}$$

Theorem 7. Every closed sphere in a metric space is a closed set.

Proof. Let $S_r[x]$ be a closed sphere in a metric space (X, d). Let $y \in X - S_r[x]$. Then $y \notin S_r[x]$ and so $d(x, y) > r$. Put $r_1 = d(x, y) - r$. Then $r_1 > 0$. Let $z \in S_{r_1}(y)$. Then $d(y, z) < r_1$. We have

$$d(x, y) \leq d(x, z) + d(y, z)$$

Therefore

$$d(x, z) \geq d(x, y) - d(y, z) > d(x, y) - r_1 = r$$

Hence $z \notin S_r[x]$ and so $z \in X - S_r[x]$. Thus

$$S_n(y) \subset X - S_r[x]$$

that is $X - S_r[x]$ is open. Hence $S_r[x]$ is closed.

Definition 10. Let $S \subset (x, d)$. The *closure* $\bar{S}$ of S is the smallest closed set containing S, i.e. $\bar{S} = \underset{F \supset S}{\cap} F \supset S$. Note that

$$\bar{S} = S \cup S' \text{ that is } x \in \bar{S} \text{ iff } S_r(x) \cap S \neq \emptyset \text{ for every } r > 0.$$

Example 21. (a) Let $\mathbb{R}$ be the usual metric space and if $S = [a, b]$, $(a, b]$, (a, b) or $[a, b)$, then $\bar{S} = [a, b]$.

(b) If $S = \mathbb{R}$, then $\bar{S} = \mathbb{R}$.

(c) If $S = \mathbb{Q}$, then $\bar{S} = \mathbb{R}$.

Remark. Let A be a subset of a metric space X. Then

(i) $\bar{\emptyset} = \emptyset$ (ii) $\bar{X} = X$ (iii) $\bar{\bar{A}} = \bar{A}$

Theorem 8. Let (X, d) be a metric space and $A \subset X$. Then

(i) $\bar{A}$ is a closed set

(ii) A is closed iff $\bar{A} = A$

(iii) $\bar{A}$ is the smallest closed subset of X containing A

(iv) $\bar{A}$ is the intersection of all closed subsets of X containing A

Proof. (i) Let x be a limit point of $\bar{A}$. Then, for a given $\varepsilon > 0$, there exists $y \in \bar{A}$ such that $d(x, y) < \varepsilon/2$. Further, since $y \in \bar{A}$, i.e. either $y \in A$ or y is a limit point of A, there exists $z \in A$ such that $d(y, z) < \varepsilon/2$. Now

$$d(x, z) \leq d(x, y) + d(y, z) < \varepsilon/2 + \varepsilon/2 = \varepsilon$$

Hence x is a limit point of A, and so $x \in \bar{A}$ This shows that $\bar{A}$ is closed.

(ii) If $\bar{A} = A$, then by (i) $\bar{A}$ is closed, and so is A. Conversely, let A be any closed set. Since $A \subseteq \bar{A}$, we need to show that $\bar{A} \subseteq A$. Let x be any element of $\bar{A}$. Then either $x \in A$ or $x \notin A$. If $x \in A$, then the result is proved. If $x \notin A$ and $x \in \bar{A}$, then for every $r > 0$, the open sphere $S_r(x)$ contains a point of A other than x, and so x is a limit point of A. But A being closed, therefore x must be in A. Hence $\bar{A} \subseteq A$.

(iii) Let B be any closed subset of X with $A \subset B$ and let $x \in \overline{A}$. If $x \in A$, then $x \in B$ and so $\overline{A} \subset B$. If $x \notin A$, then x is a limit point of A. Therefore, for a given $\varepsilon > 0$, there exists $y \in A$ such that $d(x, y) < \varepsilon$. But $y \in B$ and $A \subset B$, therefore x is also a limit point of B, since B is closed and $x \in B$, $\overline{A} \subset B$. Hence, $B \supset \overline{A} \supset A$ and $\overline{A}$ is closed by (i), we get (iii).

(iv) Let $J = \cap \, [B \subset X : B$ is closed and $B \supset A\}$. Then J is closed. Clearly J is the smallest closed subset of X containing A. Therefore, by (iii), $\overline{A} = J$, i.e. (iv).

Definition 11. Let $A, B \subset (X, d)$. Define

$$d(x, A) = \inf\{d(x, a) : a \in A\}$$
$$d(A, B) = \inf\{d(a, b) : a \in A, b \in B\}$$
$$d(A) = \sup \{d(a, a') : a, a' \in A\}$$

We call $d(x, A)$ the distance between the point $x \in X$ and the set A, $d(A, B)$ the distance between the sets A and B and $d(A)$ the diameter of the set A.

Definition 12. A set A in a metric space (X, d) is called *bounded* iff $d(A) < \infty$. Othewise it is unbounded. By convention $d(\emptyset) = -\infty$.

A metric d on a non-empty set X is said to be *bounded* if there exists a real number $k > 0$ such that $d(x, y) \leq k$, $\forall \, x, y \in X$, i.e., $d(X) \leq k$. Then (X, d) is called a *bounded metric space*, otherwise unbounded.

Example 22. (i) If A is finite, then $d(A) = \max\{d(a, a') : a, a' \in A\}$.

(ii) If $A = (0, 1) \subset \mathbb{R}$, then $d(A) = 1$.

(iii) The intervals $[a, b]$, $(a, b]$, $[a, b)$ and (a, b) are bounded in $\mathbb{R}$, while $[a, \infty)$ and $(-\infty, a]$ are not bounded.

(iv) Any finite subset in a metric space is bounded.

(v) Any set with a discrete metric is a bounded metric space.

2.5 Subspaces

We have examined several examples of metric spaces. These examples by no means exhaust the supply of known metric spaces. If we are given some metric spaces, there are several ways of constructing new metric spaces. This section discusses one of such methods, namely, subspaces.

Definition 14. Let (X, d) be a metric space and let $A \subset X$. Then the restriction map d_A of the metric d to $A \times A$ is a metric for A called *induced metric* and the metric space (A, d_A) is called a *subspace* of (X, d).

Example 23. (i) $\mathbb{Q}$ is a subspace of the usual metric space $\mathbb{R}$, and $\mathbb{R}$ is a subspace of the usual metric space $\mathbb{C}$.

(ii) The space c_0 is subspace of c and c is subspace of ℓ_∞ with the usual metric $d(x, y) = \sup|x_n - y_n|$.

(iii) The sequence spaces $^n c$, c_0, and l_∞ are subspaces of w.

(iv) The set $P[a, b]$ of all polynomials on $[a, b]$ is a subspace of the metric space $C[a, b]$ with the uniform metric $d_\infty(f, g) = \max_{a \le t \le b} |f(t) - g(t)|$.

Theorem 9. Let Y be a subspace of a metric space (X, d) and $A \subset Y$. Then

(i) A is open in Y iff there exists an open set G in X such that $A = G \cap Y$.

(ii) A is closed in Y iff there exists a closed set F in X such that $A = F \cap Y$.

For the proof of our theorem we will use the following lemma:

Lemma 10. Let Y be a subspace of a metric space (X, d) and if $a \in Y$ and $r > 0$, then

$$\tilde{S}_r(a) = Y \cap S_r(a),$$

where $S_r(a)$ and $\tilde{S}_r(a)$ are open spheres in X and Y, respectively.

Proof of the theorem. (i) Let $A = G \cap Y$ and $x \in A$ be arbitrary. Then $x \in G$ and $x \in Y$. Since G is open in X, there exists $r > 0$ such that $S_r(x) \subset G$. Also, since $x \in Y$, we get by Lemma 10

$$\tilde{S}_r(x) = Y \cap S_r(x) \subset Y \cap G = A$$

Therefore x is an interior point of a subset A of Y. Hence $A^0 = A$ in Y, i.e. A is open in Y.

Conversely, let A be open in Y and $x \in A$ be arbitrary. Then there exists an open sphere $\tilde{S}_{r_x}(x)$ such that $\tilde{S}_{r_x}(x) \subset A$.

Now

$$A = \bigcup_{x \in A} \tilde{S}_{r_x}(x) = \left(\bigcup_{x \in A} S_{r_x}(x) \right) \cap Y = G \cap Y$$

$$G = \bigcup_{x \in A} S_{r_x}(x)$$

Hence $A = G \cap Y$, where G being an arbitrary union of open spheres in X is open in X.

(ii) We know that A is closed in Y iff $Y - A$ is open in Y iff $Y - A = G \cap Y$, where G is open in X.

Therefore $A = Y - G \cap Y$, i.e. A is closed iff

$$A = Y - G \cap Y = X \cap Y - G \cap Y = (X - G) \cap Y = F \cap Y$$

where $F = X - G$ is closed.

Corollary 11. Let Y be a subspace of a metric space (X, d) and $A \subset Y$. Then

(i) if A is open in Y and Y is open in X, then A is open in X.

(ii) if A is closed in Y and Y is closed in X, then A is closed in X.

2.6 Miscellaneous Examples

1. The set $\mathbb{R}^n$ with d' defined by $d'(x, y) = \sum_{i=1}^{n} |x_i - y_i|$ is a metric space. Note that such a metric d' is called *rectangular metric* on $\mathbb{R}^n$.
We will show the triangle inequality only. Consider

$$d'(x, y) = \sum_{i=1}^{n} |x_i - y_i| = \sum_{i=1}^{n} |x_i - z_i + z_i - y_i|$$

$$\leq \sum_{i=1}^{n} |x_i - z_i| + \sum_{i=1}^{n} |z_i - y_i| = d'(x, z) + d'(z, y), \quad \forall x, y, z \in \mathbb{R}^n.$$

2. The set $C[a, b]$ is a metric space with the metric

$$d(f, g) = \left(\int_a^b (f(x) - g(x))^2 \, dx \right)^{1/2}, \quad \forall f, g \in C[a, b].$$

To establish triangle inequality, consider

$$\phi(t) = \int_a^b (tf(x) - g(x))^2 \, dx$$

$$= t^2 \int_a^b f^2(x) dx + 2t \int_a^b f(x) g(x) dx + \int_a^b g^2(x) dx, \quad t \in [a, b]$$

Since $\phi(t) \geqslant 0$, $\forall t \in [a, b]$ the discriminant of the quadratic equation in t should be non-positive, and so

$$\left(\int_a^b f(x) g(x) dx \right)^2 \leq \int_a^b f^2(x) dx \int_a^b g^2(x) dx,$$

$$\int_a^b f(x) g(x) dx \leq \left(\int_a^b f^2(x) dx \right)^{1/2} \left(\int_a^b g^2(x) dx \right)^{1/2} \quad (1)$$

Now consider

$$\left[\left(\int_a^b (f(x) - h(x))^2\, dx\right)^{\frac{1}{2}} + \left(\int_a^b (h(x) - g(x))^2\, dx\right)^{\frac{1}{2}}\right]^2$$

$$= \int_a^b (f(x) - h(x))^2\, dx + \int_a^b (h(x) - g(x))^2\, dx$$

$$+ 2\left(\int_a^b (f(x) - h(x))^2\, dx\right)^{\frac{1}{2}} \left(\int_a^b (h(x) - g(x))^2\, dx\right)^{\frac{1}{2}}$$

$$\geq \int_a^b (f(x) - h(x))^2\, dx + \int_a^b (h(x) - g(x))^2\, dx$$

$$+ 2\int_a^b (f(x) - h(x))(h(x) - g(x))dx, \quad \text{by (1)}$$

$$= \int_a^b (f(x) - h(x) + (h(x) - g(x))^2\, dx$$

$$= \int_a^b (f(x) - g(x))^2\, dx$$

Hence

$$d(f, g) \leq d(f, h) + d(h, g), \quad \forall\, f, g\, h \in C[a, b]$$

3. Let (X, d) be a metric space. Then the function defined by

$$d_1(x, y) = \frac{d(x, y)}{1 + d(x, y)}$$

is also a metric on X.

Since d is a metric on X, we have, for all $z \in X$

$$d(x, y) \leq d(x, z) + d(z, y)$$

or

$$1 + d(x, y) \leq 1 + d(x, z) + d(z, y)$$

or

$$1 - \frac{1}{1 + d(x, y)} \leq 1 - \frac{1}{1 + d(x, z) + d(z, y)}$$

or

$$\frac{d(x, y)}{1 + d(x, y)} \leq \frac{d(x, z) + d(z, y)}{1 + d(x, z) + d(z, y)}$$

$$\leq \frac{d(x, z)}{1 + d(x, z)} + \frac{d(z, y)}{1 + d(z, y)}$$

i.e., $\qquad d_1(x, y) \le d_1(x, z) + d_1(z, y)$

4. Let $\ell_2 = \left\{ x = \{x_n\} \in w: \sum\limits_{n=1}^{\infty} |x_n|^2 < \infty \right\}$

(ℓ_2, d) is a metric space with the metric d defined by

$$d(x, y) = \left(\sum_{n=1}^{\infty} (x_n - y_n)^2 \right)^{1/2}, \quad \forall\ x = (x_n),\ y = (y_n) \in \ell_2$$

To establish triangle inequality, let $x = (x_n)$, $y = (y_n)$ and $z = (z_n) \in \ell_2$. Then

$$\left(\sum_{n=1}^{k} (x_n - y_n)^2 \right)^{1/2} = \left(\sum_{n=1}^{k} (x_n - z_n + z_n - y_n)^2 \right)^{1/2} \quad \text{for all } k$$

$$\le \left(\sum_{n=1}^{k} (x_n - z_n)^2 \right)^{1/2} + \left(\sum_{n=1}^{k} (z_n - y_n)^2 \right)^{1/2}$$

by the Minkowski's inequality.
Letting $k \to \infty$, we get

$$\left(\sum_{n=1}^{\infty} (x_n - z_n)^2 \right)^{1/2} \le \left(\sum_{n=1}^{\infty} (x_n - z_n)^2 \right)^{1/2} + \left(\sum_{n=1}^{\infty} (z_n - y_n)^2 \right)^{1/2}$$

Hence $d(x, y) \le d(x, z) + d(z, y)$.

Other requirements are easily verified.

5. Let I be the set of all integral functions, i.e. functions f which are analytic for all finite z, e.g. polynomials, e^z, $\sin z$, etc. If we write

$$M_n = \max_{|z|=n} |f(z) - g(z)|, \quad n = 1, 2, \ldots$$

Then $\qquad d(f, g) = \sum\limits_{n=1}^{\infty} \dfrac{1}{2^n} \dfrac{M_n}{1 + M_n}$ is a metric on I.

This can similarly be verified as in case of w.

6. Every set in a discrete metric space (X, d) is open.

Let G be any non-empty subset of the discrete metric space (X, d) and $x \in G$. Then the open sphere $S_r(x)$ with $r \le 1$ is the singleton set $\{x\} \subseteq G$, that is, each point of G is the centre of some open sphere contained in G. In particular each singleton set is open.

7. In $\mathbb{R}$ with the usual metric, the singleton set $\{x\}$ is not open.

For the metric space $(\mathbb{R}, d)$ each open sphere $S_r(x)$ is the bounded open interval $(x - r, x + r)$ and for no value of r this sphere is contained in $\{x\}$. Hence $\{x\}$ is not open in $(\mathbb{R}, d)$.

8. Let d and d' be, respectively, the usual and discrete metrics on $\mathbb{R}$. Then every singleton set $\{x\}$, $x \in \mathbb{R}$ is open in $(\mathbb{R}, d')$ but not so in $(\mathbb{R}, d)$. Every singleton set $\{x\}$, $x \in \mathbb{R}$ is open in $(\mathbb{R}, d')$ being an open sphere $S_r(x)$, $r < 1$, i.e., bounded open interval $(x - r, x + r)$ for $r < 1$ contains only one point of the set x of the space. But $\{x\}$ is not open in $(\mathbb{R}, d)$, since every open sphere $S_r(x)$, $r < 1$ is the bounded open interval $(x - r, x + r) \not\subset \{x\}$.

9. Let $\mathbb{R}_\infty$ be the *extended set* of real numbers (i.e. the set of real numbers including $-\infty$ and $+\infty$).

The function d defined by

$$d(x, y) = |f(x) - f(y)|, \quad \forall \quad x, y \in \mathbb{R}_\infty,$$

where

$$f(x) = \begin{cases} x\big/(1 + |x|) & \text{if} \quad -\infty < x < \infty \\ 1 & \text{if} \quad x = \infty \\ -1 & \text{if} \quad x = -\infty \end{cases}$$

Then $(\mathbb{R}_\infty, d)$ is a bounded metric space.

Since

$$\left| \frac{x}{1 + |x|} - \frac{y}{1 + |y|} \right| \leq \left| \frac{x}{1 + |x|} - \frac{z}{1 + |z|} \right| + \left| \frac{z}{1 + |z|} - \frac{y}{1 + |y|} \right|,$$

we have

$$d(x, y) \leq d(x, z) + d(z, y)$$

If $x = +\infty$, $y = -\infty$, then

$$d(x, y) = |1 - (-1)| \leq \left| 1 - \frac{z}{1 + |z|} \right| + \left| \frac{z}{1 + |z|} - (-1) \right|$$

$$= d(x, z) + d(z, y)$$

Similarly, when $x = -\infty$, $y = +\infty$, the triangle inequality holds.

Moreover, if x and y are two elements of $\mathbb{R}_\infty$, then

$$-1 \leq f(x) \leq 1 \quad \text{and} \quad -1 \leq f(y) \leq 1$$

Therefore

$$d(x, y) = |f(x) - f(y)| \leq 2, \quad \forall \ x, y \ \in \mathbb{R}_\infty$$

Hence $(\mathbb{R}_\infty, d)$ is a bounded metric space.

10. The metric space c_0 is unbounded.

We can find two null sequences whose distance can be made arbitrary large so that $d(c_0) = \infty$. For example $e_0 = (0, 0, 0, ...)$ and $e(n) = (n, 0, 0, ...) \in c_0$ for which $d(e_0, e(n)) = n$, so that $d(c_0) = \infty$.

EXERCISES

1. Verify that each of the following is a metric on $\mathbb{R}^2$

   $$d(x, y) = [(x_1 - y_1)^2 + (x_2 - y_2)^2]^{1/2}$$
   $$d(x, y) = |x_1 - y_1| + |x_2 - y_2|$$
   $$d_\infty(x, y) = \max\{|x_1 - y_1|, |x_2 - y_2|\}$$

2. Let cs: $= \{x = (x_k): \sum\limits_k x_k$ is convergent$\}$, the space of convergent series; and let

   $$d(x, y) = \sup_n \left| \sum_{k=1}^n (x_k - y_k) \right|, \, x = (x_k), \, y = (y_k) \in \text{cs}$$

   Show that (cs, d) is a metric space.

3. Prove that the function $d(x, y) = \sum\limits_n |x_n - y_n|^p$ defines a metric on ℓ_p for $0 < p < 1$.

4. Let S be a non-empty set and let $X = B(s)$ denote the collection of all bounded real valued functions defined on S. Show that

   $$d(f, g) = \sup\{|f(t) - g(t)|: t \in S\}$$

   is a metric on S.

5. Show that the spaces $\ell_\infty(p)$, $c_0(p)$ and $c(p)$ are all metric spaces with the metric

   $$d(x, y) = \sup_k |x_k - y_k|^{p_k/M}$$

   where $M = \max\{1, H\}$, $H = \sup\limits_k p_k < \infty$.

6. Prove that every open set in the usual metric space $\mathbb{R}$ is the union of a countable disjoint class of open intervals.

7. Prove that the set $\mathbb{R}^n$ of n-tubles is a metric space with respect to each of the following metrics:

   (i) $$d(x, y) = \left(\sum_{i=1}^n (x_i - y_i)^2 \right)^{1/2}$$

   (ii) $$d_1(x, y) = \max_{1 \le i \le n} |x_i - y_i|$$

   for all $x = (x_i) = (x_1, x_2, ..., x_n)$,

   $$y = (y_i) = (y_1, y_2, ..., y_n) \in \mathbb{R}^n.$$

   Also prove that

   $$d_1(x, y) \le d(x, y) \le \sqrt{n}\, d_1(x, y)$$

8. Show that $(\mathbb{C}, d)$ is a metric space, where

   $$d(x, y) = \begin{cases} |x| + |y|, & \text{if } x \ne y, \\ 0 & \text{if } x = y. \end{cases}$$

9. Show that the conditions (M_2) and (M_3) of Definition 1 are not sufficient to ensure that the function $d: X \times X \to \mathbb{R}$ is a metric on a non-empty set X.

10. Let $H_\infty = \{x = (x_n) \in w: |x_n| \le 1, \ \forall \ n \in \mathbb{N}\}$. Prove that (H_∞, d) is a metric space,

   where $d(x, y) = \sum\limits_{n=1}^{\infty} \dfrac{1}{2^n} |x_n - y_n|, \ \forall \ x = (x_n), y = (y_n) \in H_\infty$.

11. Prove that the function $d: \mathbb{C} \times \mathbb{C} \to \mathbb{R}$ defined by

   $$d(x, y) = \frac{2|x - y|}{\sqrt{1 + |x|^2} \sqrt{1 + |y|^2}}$$

   is a metric on $\mathbb{C}$.

12. Let $d: \mathbb{N} \times \mathbb{N} \to \mathbb{R}$ be defined by

   $$d(m, n) = \begin{cases} 0, & \text{if } m = n, \\ \dfrac{1}{5^k}, & \text{if } m \ne n; \end{cases}$$

   for $m, n \in \mathbb{N}$, where $m - n = 5^r$ and r is not a multiple of 3. Prove that $(\mathbb{N}, d)$ is a bounded metric space.

13. If (X, d_1) and (Y, d_2) are two metric spaces, prove that the Cartesian product

   $$X \times Y = \{(x, y): x \in X, y \in Y\}$$

   is a metric space with the following metric

   $$\sigma((x_1, y_1), (x_2, y_2)) = [d_1(x_1, x_2)^2 + d_1(y_1, y_2)^2]^{1/2}$$

14. If (X, ρ) is a metric space, prove that (X, σ) is also a metric space, where

   $$\sigma(x, y) = \min (1, \rho(x, y)).$$

15. Let A and B be any two non-empty subsets of a metric space X. Prove that

   (i) if $A \subseteq B$ then $d(A) \le d(B)$.
   (ii) $d(A \cup B) \le d(A) + d(B) + d(A, B)$.
   (iii) If $A \cap B = \emptyset$, then $d(A \cup B) \le d(A) + d(B)$.

16. Let $X = \mathbb{R}^2$, $A: = \{x = (x_1, x_2): (|x_1|^2 + |x_2|^2)^{1/2} < 1\}$, and $d(x, y) = \{|x_1 - y_1|^2 + |x_2 - y_2|^2\}^{1/2}$. Compute $d(x, A)$. Show that $d(x, A) = 0$ iff $|x_1|^2 + |x_2|^2 \le 1$.

17. Let $A, B \subset (X, d)$. Then prove that

   (a) $A \subset B \Rightarrow \bar{A} \subset \bar{B}$ (b) $\overline{A \cup B} = \bar{A} \cup \bar{B}$ (c) $\overline{A \cap B} \subseteq \bar{A} \cap \bar{B}$

18. In a metric space (X, d), prove that $\overline{S_r(a)} \subset S_r[a]$.

19. Let $A \subset (X, d)$. Prove that $\overline{X - A} = X - A^0$.

20. If $a \in \mathbb{R}$, prove that $[a, \infty)$ is a closed subset of $\mathbb{R}$.

21. Let $A = \left\{ \left(\dfrac{m}{n}, \dfrac{1}{n} \right) : n = 1, 2, \ldots; m = 0, \pm 1, \pm 2, \ldots \right\} \subseteq \mathbb{R}^2$. Prove that

$$\overline{A} = A \cup \{(x, 0): x \in \mathbb{R}\}.$$

22. Find the frontier of the subset $\{(x_1, x_2): x_2 = 0\}$ of $\mathbb{R}^2$.
23. Let A and B be any two subsets of a metric space X. Prove that

    (i) ext (A) is the largest open set contained in A^c.
    (ii) A^c is open iff $A^c = $ ext (A).
    (iii) $A \subseteq B \Rightarrow $ ext $(A) \supseteq $ ext (B).
    (iv) ext $(A \cap B) \supseteq $ ext $(A) \cup $ ext (B).
    (v) ext $(A \cup B) = $ ext $(A) \cap $ ext (B).

24. Let A and B be two subsets of a metric space X. Prove that

    (i) $F_r(A) = \overline{A} \cap (\overline{A})^c = \overline{A} - A^0$
    (ii) $F_r(A) = \varnothing$ iff A is both open and closed
    (iii) A is closed iff $A \supseteq F_r(A)$
    (iv) A is open iff $A^c \supseteq F_r(A)$
    (v) $F_r(A \cap B) \subseteq F_r(A) \cup F_r(B)$. The equality holds if $\overline{A} \cap \overline{B} = \varnothing$
    (vi) $F_r(A^0) \subseteq F_r(A)$

25. Prove that an open sphere in a metric space is a bounded set.
26. Prove that any subset of a discrete metric space has no limit point.
27. Prove that the sequence space cs defined in Exercise 2 is a subspace of the following sequence space

$$bs = \left\{ x = (x_k): \sup_n \left| \sum_{k=1}^{n} x_k \right| < \infty \right\}$$

with the metric $d(x, y)$ as given in Exercise 2.

28. (The Closed Set Theorem). Prove that a set A in a metric space (X, d) is a closed set iff every convergent sequence $\{x_n\}$ with $\{x_n\} \subset A$ has its limit in A.

3

Completeness

Consider the metric space $X := \{x \in \mathbb{R} : 0 < x \leq 1\}$ with the usual metric $d(x, y) = |x - y|$. The sequence $\{1/n\} = \{1, 1/2, 1/3, \ldots\} \in X$. At first glance we may say that this sequence is convergent to 0. However, $0 \notin X$. Therefore, to say that 0 is the limit is not justified. This particular sequence in X is just not convergent. On the other hand, we still feel that there is something special about this sequence and something different about the manner in which it fails to have a limit. What is important is that the sequence is a "Cauchy sequence" and the difference is that the metric X is not complete. Hence, we see that a Cauchy sequence in a metric space need not to be convergent while in $\mathbb{R}$ (or $\mathbb{C}$), every Cauchy sequence is also convergent, and it is because of the completeness of $\mathbb{R}$ (or $\mathbb{C}$).

3.1 Convergent and Cauchy Sequences

First let us look at the analogue definitions of convergent and Cauchy sequences in a metric space.

Definition 1. A sequence $\{x_n\}$, where $x_n \in X$ for every n, is said to be a *Cauchy sequence* in a metric space (X, d) if for every $\varepsilon > 0$ there exists an $N = N(\varepsilon)$ such that $d(x_n, x_m) < \varepsilon$ for every choice of $n, m \geq N$. Another way of defining is $d(x_n, x_m) \to 0$ as $m, n \to \infty$.

Definition 2. A sequence $\{x_n\}$ in a metric space (X, d) is said to be *convergent* to x if there exists $x \in X$ such that $d(x_n, x) \to 0$ as $n \to \infty$, then we write $x = \lim x_n$ or $x_n \to x$ and call x the limit of the sequence $\{x_n\}$. A sequence which is not convergent is said to be *divergent*.

The first theorem is just an analogue of convergent sequences in $\mathbb{R}$ (or $\mathbb{C}$).

Theorem 1. In a metric space every convergent sequence has a unique limit.

Proof. Let, if possible, the sequence $\{x_n\}$ converge to x and y, i.e. $d(x_n, x) \to 0$ and $d(x_n, y) \to 0$ as $n \to \infty$. Then by the triangle inequality

$$0 \le d(x. y) \le d(x, x_n) + d(x_n, y) \to 0 \ (n \to \infty)$$

Hence $d(x, y) = 0$ by (M2) we have $x = y$, i.e. the limit is unique. Note that if d is merely a semi-metric, then we cannot conclude from $d(x, y) = 0$ that $x = y$.

The following theorem presents the key connection between convergent sequences and Cauchy sequences.

Theorem 2. (i) Every convergent sequence is also a Cauchy sequence, but not conversely, in general.

(ii) If a Cauchy sequence has a convergent subsequence then the whole sequence is convergent.

Proof. (i) Let $\{x_n\}$ be convergent to x. Then $d(x_n, x) \to 0$ as $n \to \infty$, so

$$0 \le d(x_n, x_m) \le d(x_n, x) + d(x, x_m) \to 0 \text{ as } n, m \to \infty$$

Hence $\{x_n\}$ is Cauchy. For the converse part, as earlier, consider the sequence $\{1/n\}$ in the usual metric space $X := \{x \in \mathbb{R} : 0 < x \le 1\}$. Then

$$d(x_n, x_m) = \left| \frac{1}{n} - \frac{1}{m} \right| \le \frac{1}{n} + \frac{1}{m} \to 0 \ (n, m \to \infty)$$

i.e. $\{1/n\}$ is Cauchy in X but $\{1/n\}$ is trying to converge to 0, which is not in X. More precisely, we have that $\frac{1}{n} \to 0$ in $\mathbb{R}$ and that 0 is the unique limit so that no point of $(0, 1]$ could also be the limit.

(ii) Let $\{x_n\}$ be Cauchy, i.e. $d(x_n, x_m) \to 0$ $(n, m \to \infty)$. Suppose that $\{x_{n_k}\}$ is a convergent subsequence, i.e. $d(x_{n_k}, x) \to 0$ $(k \to \infty)$. Thus, we have

$$0 \le d(x_n, x) \le d(x_n, x_{n_k}) + d(x_{n_k}, x) \to 0$$

so that Cauchy sequence $\{x_n\}$ converges to the limit of the convergent subsequence $\{x_{n_k}\}$.

Remark. If a subsequence of a sequence in (X, d) is convergent, then the sequence itself need not be convergent, e.g. consider the sequence $\{x_n\} = \{(-1)^n\}$ in the usual metric space $\mathbb{R}$. We know that $\{x_n\}$ is not convergent but its subsequence $\{x_{2n}\} = \{1, 1, 1, \ldots\}$ is convergent to 1.

3.2 Complete Metric Spaces

We have seen that every convergent sequence is a Cauchy sequence but a Cauchy sequence need not be convergent. The above example suggests that the reason some Cauchy sequences fail to converge is a fault of the underlying set X. In this example, if we replace $(0, 1]$ by $[0, 1]$, then the sequence $\{1/n\}$ would converge in $X = [0, 1]$, i.e. in the larger space a Cauchy sequence is also convergent. Metric spaces possessing this property are called *complete metric spaces*.

Definition 3. A metric space (X, d) is said to be *complete* if each Cauchy sequence in (X, d) is a convergent sequence in (X, d).

Example 1. (i). The set $\mathbb{Q}$ of rationals with usual metric is not a complete metric space.

Consider, e.g. the sequence $\{3, 3.1, 3.14, 3.141, 3.1415, \ldots\}$ in $\mathbb{Q}$. This sequence is Cauchy but not convergent in $\mathbb{Q}$, since $\pi \notin \mathbb{Q}$.
(ii) $\mathbb{R}$ and $\mathbb{C}$ are complete with the usual metrics.
(iii) The discrete metric space X_d is complete.

For if $\{x_n\}$ is Cauchy in X_d then, on taking $\varepsilon = 1/2$ in the Cauchy sequence definition, we have $d(x_n, x_{N+1}) < 1/2 \ \forall \ n > N = N(1/2)$ and so $d(x_n, x_{N+1}) = 0$ for such n, whence it is clear that $x_n \to X_{N+1}$ as $n \to \infty$. Thus, every Cauchy sequence converges to a point in X_d, actually a member of the sequence in this case. Hence, X_d is complete.

Example 2. The set $\mathbb{C}^n$ is complete with the metric

$$d(x, y) = \left(\sum_{k=1}^{n} |\xi_k - \eta_k|^2 \right)^{1/2}, \text{ where } x = (\xi_1, \xi_2, \ldots, \xi_n),$$

$$y = (\eta_1, \eta_2, \ldots, \eta_n) \in \mathbb{C}^n$$

Take a Cauchy sequence $\{x_m\}_1^\infty$ in $\mathbb{C}^n$, $x_m \in \mathbb{C}^n \ \forall \ m$.

Let $\quad x_m = (\xi_1^m, \xi_2^m, \ldots, \xi_n^m), x_p = (\xi_1^p, \xi_2^p, \ldots \xi_n^p)$

Then $\quad\quad\quad\quad d(x_m, x_p) < \varepsilon \ (m, p \geq N)$

$\Rightarrow \quad \left(\sum_{l=1}^{n} |\xi_l^m - \xi_l^p|^2 \right)^{1/2} < \varepsilon \Rightarrow \sum_{l=1}^{n} |\xi_l^m - \xi_l^p|^2 < \varepsilon^2$

$\Rightarrow \quad |\xi_l^m - \xi_l^p|^2 < \varepsilon^2 \quad \text{or} \quad |\xi_l^m - \xi_l^p| < \varepsilon \ (m, p \geq N)$

for a fixed l. Hence $\{\xi_l^m\}$ is a Cauchy sequence in $\mathbb{C}$. Since $\mathbb{C}$ is complete, $\{\xi_l^m\}$ is convergent to ξ_l in $\mathbb{C}$, i.e.

$$\{\xi_l^m\} \to x = (\xi_1, \xi_2, \ldots, \xi_n) \in \mathbb{C}^n$$

Hence $\mathbb{C}^n$ is complete.

Example 3. The metric space $X = C[0, 1]$ is
(i) Complete with the metric

$$d(f, g) = \max \{|f(x) - g(x)| : 0 \le x \le 1\}$$

(ii) Not complete with the metric

$$d_1(f, g) = \int_0^1 |f(x) - g(x)| \, dx$$

(i) Let $\{f_m(x)\}_1^\infty$ be a Cauchy sequence in X. Then

$$d(f_m, f_n) < \varepsilon \ (m, n \ge N), \text{ i.e. } \max_{0 \le x \le 1} |f_m(x) - f_n(x)| < \varepsilon$$

$$\Rightarrow \quad |f_m(x) - f_n(x)| < \varepsilon \Rightarrow \{f_m(x)\} \text{ converges uniformly}$$

Since each $f_m(x)$ is a continuous function on $[0, 1]$, and we have shown that $\{f_m\}$ is uniformly convergent; so that its limit, say, f is also continuous on $[0, 1]$, i.e.

$$f_m(x) \to f(x) \in C[0, 1]$$

Therefore

$$\sup_{0 \le x \le 1} |f_m(x) - f(x)| \to 0 \ (m \to \infty), \text{ i.e. } d(f_m, f) < \varepsilon \ (m \ge N)$$

Hence $(C[0, 1], d)$ is complete.
(ii) Let us assume $f_n \to f$ in $(C[0, 1]; d_1)$. Then, we can show that f cannot be continuous at $1/2$. Consider

$$f_n(x) = \begin{cases} 1, & 0 \le x \le \dfrac{1}{2} \\ 1 - 2^n\left(x - \dfrac{1}{2}\right), & \dfrac{1}{2} < x \le \dfrac{1}{2} + \dfrac{1}{2^n} \\ 0, & \dfrac{1}{2} + \dfrac{1}{2^n} < x \le 1 \end{cases}$$

Then $f_n(x) - f_{n+1}(x) = 1 - 2^n\left(x - \dfrac{1}{2}\right) - \left[1 - 2^{n+1}\left(x - \dfrac{1}{2}\right)\right]$

$$= 1 - 2^n\left(x - \frac{1}{2}\right) - 1 + 2^{n+1}\left(x - \frac{1}{2}\right)$$

$$= (-2^n + 2^{n+1})\left(x - \frac{1}{2}\right) = 2^n(-1 + 2)\left(x - \frac{1}{2}\right)$$

$$= 2^n\left(x - \frac{1}{2}\right)$$

and so

$$d_1(f_n(x), f_{n+1}(x)) = \int_0^{1/2} (1-1)\, dx + \int_{\frac{1}{2}}^{\frac{1}{2}+\frac{1}{2^n}} 2^n \left(x - \frac{1}{2}\right) dx + \int_{\frac{1}{2}+\frac{1}{2^n}}^1 0\, dx$$

$$= 2^n \int_{\frac{1}{2}}^{\frac{1}{2}+\frac{1}{2^n}} \left(x - \frac{1}{2}\right) dx$$

$$= 2^n \left[\frac{(x - 1/2)^2}{2}\right]_{\frac{1}{2}}^{\frac{1}{2}+\frac{1}{2^n}} = 2^n \left[\frac{1}{2^{2n}} \cdot \frac{1}{2}\right]$$

$$= \frac{1}{2} \cdot \frac{1}{2^n} \to 0 \text{ as } n \to \infty$$

Hence $\{f_n(x)\}$ is a Cauchy sequence in $C[0, 1]$. But

$$\lim_{n \to \infty} f_n(x) = f(x) = \begin{cases} 1, & 0 \le x \le \frac{1}{2} \\ 0, & \frac{1}{2} < x \le 1 \end{cases}$$

which is not continuous at $x = 1/2$, i.e. the limit function of $\{f_n(x)\}$ does not belong to $C[0, 1]$. Therefore, $(C[0, 1], d_1)$ is not complete.

Example 4. The space $l(p)$ is complete with the metric defined by

$$d(x, y) = \left(\sum_{k=1}^\infty |x_k - y_k|^{p_k}\right)^{1/M}$$

$\forall\, x = (x_k),\, y = (y_k) \in l_p$, where $M = \max\{1, H\}$, $H = \sup_k p_k < \infty$.

Let $\{z^n\}$ be a Cauchy sequence in $l(p)$ then $d(z^n, z^m) \to 0\, (n, m \to \infty)$
$\Rightarrow d(z^n, z^m) < \varepsilon,\, n, m \ge N$. Therefore

$$\sum_k |z_k^n - z_k^m|^{p_k} < \varepsilon,\, n, m \ge N$$

$$\Rightarrow \qquad |z_k^n - z_k^m|^{p_k} < \varepsilon,\, n, m \ge N$$

so that $|z_k^n - z_k^m| < \varepsilon,\, n, m \ge N$. Hence, for each k the sequence $(z_k^m) = (z_k^1, z_k^2, \ldots)$ is Cauchy in $\mathbb{R}$ (or $\mathbb{C}$). Let $z_k^m \to z_k\, (m \to \infty)$ for each k. Then

$$|z_k^n - z_k| < \varepsilon\, (n \ge N) \quad \Rightarrow \quad \left(\sum_k |z_k^n - z_k|^{p_k}\right)^{1/M} < \varepsilon\, (n \ge N)$$

Therefore, $d(z^n, z) \to 0\, (n \to \infty)$, where $z = (z_k)$. Now, since

$$|z_k| \le |z_k - z_k^n| + |z_k^n|, \quad \text{so} \quad \sum_k |z_k|^{p_k} < \infty$$

Hence $z \in l(p)$, and $(l(p), d)$ is complete.

Example 5. Space $c(p)$ is complete with $d(x, y) = \sup_k |x_k - y_k|^{p_k/M}$

Let $\{x^i\}$ be a Cauchy sequence in $c(p)$. Then for each k, $\{x_k^i\}$ is a Cauchy sequence of complex numbers and hence $x_k^i \to x_k$ for each k. Now, for a given $\varepsilon > 0$ there is n_0 such that $|x_k^i - x_k^j|^{p_k/M} < \dfrac{\varepsilon}{5}$ for $i, j > n_0$. Letting $j \to \infty$, we get $|x_k^i - x_k|^{p_k/M} \le \dfrac{\varepsilon}{5}$ and therefore $x^i \to x$. We have to show that $x \in c(p)$. Since $x^i \in c(p)$ there exists $l^i \in \mathbb{C}$ such that

$$|x_k^i - l^i|^{p_k/M} < \varepsilon/5$$

Therefore

$$|l^i - l^j|^{p_k/M} \le |x_k^i - x_k^j|^{p_k/M} + |x_k^i - l^i|^{p_k/M} + |x_k^j - l^j|^{p_k/M} < \frac{3\varepsilon}{5}$$

Thus $\{l^i\}$ is a Cauchy sequence in $\mathbb{C}$ and hence there exists l such that

$$|l^i - l|^{p_k/M} < \frac{3\varepsilon}{5}$$

Therefore

$$|x_k - l|^{p_k/M} \le |x_k^i - x_k|^{p_k/M} + |x_k^i - l^i|^{p_k/M} + |l^i - l|^{p_k/M}$$

$$< \frac{\varepsilon}{5} + \frac{\varepsilon}{5} + \frac{3\varepsilon}{5} = \varepsilon$$

Hence $c(p)$ is complete.

3.3 Some Properties of Completeness

Let (X, d) be a metric space and $Y \subset X$, then (Y, d) is also a metric space. We seek conditions under which the subspace (Y, d) is complete.

Theorem 3. Let (X, d) be a complete metric space and let (Y, d) be a subspace of (X, d). Then (Y, d) is complete iff Y is a closed set in (X, d).

Proof. Let (Y, d) be complete. In order to show that Y is closed in (X, d), we will show that Y contains all of its limit points. If y is a limit point of Y, then each open ball $B_{1/n}(y)$, $n = 1, 2, \ldots$ contains a point y_n in Y. Since $d(y_n, y) < 1/n$, $\{y_n\}$ is a convergent sequence in (X, d) converging to y. However, the sequence $\{y_n\}$ is a Cauchy sequence in the complete metric space (Y, d). Therefore, $\{y_n\}$ converges to a point y_0 in (Y, d). Since the limit of a sequence is unique, $y = y_0$ or y is in Y. Hence Y is closed set in (X, d).

Conversely, let $\{y_n\}$ be a Cauchy sequence of points in (Y, d). We have to show that $\{y_n\}$ converges to a point in Y. Since $Y \subset X$, $\{y_n\}$ is a Cauchy sequence in X. Thus since X in complete, $\{y_n\}$ must converge to a point y in X, i.e.

$$\{y_n\} \in Y \quad \text{and} \quad y_n \to y \, (n \to \infty)$$

$$\Rightarrow \qquad y \in \overline{Y} = Y, \text{ as } Y \text{ is closed}$$

Hence Y is complete.

Remark. Note that Y is always a closed set in (Y, d) but this does not mean that Y is a closed set in (X, d). We make use of the above theorem for the following example.

Example 6. (i) Let d_∞ be the sup-metric on $C[0, T]$. Let $P[0, T]$ be the subset of $C[0, T]$ made up of all polynomials in t. The metric space $(C[0, T], d_\infty)$ is complete. However, the subset $P[0, T]$ is not closed. For example, the sequence in $P[0, T]$ given by

$$\left\{ 1, 1 + t, 1 + t + \frac{1}{2!}t^2, 1 + t + \frac{1}{2!}t^2 + \frac{1}{3!}t^3, \ldots \right\}$$

converges to $e^t \notin P[0, T]$. Since $P[0, T]$ is not closed, the subspace $(P[0, T], d_\infty)$ is not complete.

(ii) $[0, 1]$ is a closed subspace of $\mathbb{R}$. Since $\mathbb{R}$ is complete with the usual metric d, $([0, 1], d)$ is also complete.

Definition 4. Let X_1 and X_2 be two metric spaces. Then $f : X_1 \to X_2$ is called a *homeomorphism* if f is bijective and both f and f^{-1} are continuous on X_1 and X_2, respectively. If a *homeomorphism* $f : X_1 \to X_2$ exists, then X_1 and X_2 are said to be *homeomorphic*.

Remark. (1) If f is a *homeomorphism*, so is f^{-1}

(2) *Homeomorphism* maps open (closed) subset of X_1 onto open (closed) subset of X_2.

Example 7. (i) The entire real line is homeomorphic to an open interval. The open interval $\left(-\frac{\pi}{2}, \frac{\pi}{2} \right)$ is homeomorphic to $(-\infty, \infty)$ by the mapping $f(x) = \tan^{-1}x$. Since any two open intervals are homeomorphic, $\left(-\frac{\pi}{2}, \frac{\pi}{2} \right)$ is homeomorphic to (a, b). Hence, (a, b) is homeomorphic to $(-\infty, \infty)$.

(ii) The metric spaces $[0, 1]$ and $[a, b]$ with the usual metric are homeomorphic. A suitable homeomorphism is given by $f(x) = a + (b - a) x$.

(iii) A closed interval is not homeomorphic to an open interval of the real line.

Definition 5. Let (X, d_1) and (Y, d_2) be metric spaces. Then a mapping $f : X \to Y$ is *isometry* if it is surjective and for all $x, x' \in X$, $d_2(f(x), f(x')) = d_1(x, x')$.

Metric spaces are called *isometric* if there exists an isometry between them.

Remark. It follows that if $d_1(x, x') \neq 0$, then $d_2(f(x), f(x')) \neq 0$. Hence, $x \neq y \Rightarrow f(x) \neq f(y)$ so that if f is an isometry, then f is injective. Clearly, f is bicontinuous, in fact uniformly bicontinuous (f, f^{-1} are uniformly continuous). Hence, f is a homeomorphism.

Example 8. Let $\mathbb{R}$ be the metric space of reals and $F^* : = \{f : f(x) = \alpha x, \alpha \in \mathbb{R}\}$. If $f(x) = \alpha x$, $g(x) = \beta x$, let us define $d(f, g) = |\alpha - \beta|$. It is easy to check that (F^*, d) is a metric space. Let $T : F^* \to \mathbb{R}$ be defined by $T(f) = \alpha$. Then

$$|T(f) - T(g)| = |\alpha - \beta| = d(f, g)$$

Hence $\mathbb{R}$ and (F^*, d) are isometric

Definition 6. A property P is said to be a *topological property* if whenever a metric space (X, d) possesses P every metric space homeomorphic to (X, d) possesses property P.

Theorem 4. Let f be an isometry from (X_1, d_1) onto (X_2, d_2). Then f is a homeomorphism.

Proof. Since X_1 and X_2 are isometric, $f : X_1 \to X_2$ is bijective. Hence, $f^{-1} : X_2 \to X_1$ is well-defined. We have to show that f and f^{-1} are continuous.

Let $d_1(x_n, x) \to 0$ as $n \to \infty$ in X_1. Since f is an isometry, we have

$$d_1(x_n, x) = d_2(f(x_n), f(x))$$

so that

$$d_2(f(x_n), f(x)) \to 0 \text{ as } n \to \infty \text{ in } X_2$$

Hence, f is continuous.

Let $y_n \to y$ in X_2. Since f is bijective, there exists distinct x_n and x in X such that $f(x_n) = y_n$ and $f(x) = y$ so that $x_n = f^{-1}(y_n)$ and $x = f^{-1}(y)$. But

$$d_1(f^{-1}(y_n), f^{-1}(y)) = d_1(x_n, x) = d_2(f(x_n), f(x)) = d_2(y_n, y)$$

as f is isometry. Since $d_2(y_n, y) \to 0$ $(n \to \infty)$, we have

$$d_1(f^{-1}(y_n), f^{-1}(y)) \to 0 \ (n \to \infty) \text{ in } X_1$$

Hence f^{-1} is continuous. Therefore, f is homeomorphism.

Theorem 5. Let (X, d_1) and (Y, d_2) be two metric spaces that are isometric to one another. Then (X, d_1) is complete iff (Y, d_2) is complete, i.e. completeness is preserved by isometries.

We leave the proof as an exercise.

Although Theorem 5 is not a surprise, the next point may be somewhat of a shock.

Homeomorphism do not necessarily preserve completeness. It follows, then, that completeness is not a topological property. But how can this be since completeness has something to do with convergence and homeomorphism preserves convergence. This is because a homeomorphism does not necessarily preserve Cauchy sequence (see the following example).

Example 9. Let $X = (0, 1]$ and d_1 be the usual metric and let $Y = [1, \infty)$ and d_2 be the usual metric. The function $f : (X, d_1) \to (Y, d_2)$ defined by $y = f(x) = 1/x$, is a homeomorphism between (X, d_1) and (Y, d_2). The sequence $\{1/n\}$ in (X, d_1) is a Cauchy sequence. However, the corresponding sequence in (Y, d_2), i.e. $\{f(1/n)\} = \{n\}$ is not Cauchy in (Y, d_2).

In fact every subset of X and every subset of Y is open in the respective spaces. But (X, d_1) is a bounded metric space and (Y, d_2) is not. Also (Y, d_2) is a complete metric space but (X, d_1) is not.

Isometries then belong to the special class of homeomorphism that preserve completeness, whereas an arbitrary homeomorphism need not preserve completeness.

Definition 7. A sequence $\{A_n\}$ of subsets of a metric space (X, d) is said to be *decreasing sequence* (or *nested sequence*) of subsets if $A_1 \supseteq A_2 \supseteq A_3 \supseteq \cdots$.

Theorem 6. (*Cantor's Intersection Theorem*). Let (X, d) be a complete metric space, and let $\{F_n\}$ be a nested sequence of non-empty closed subsets of X such that $d(F_n) \to 0$ $(n \to \infty)$. Then $F = \bigcap_{n=1}^{\infty} F_n$ is non-empty and contains exactly one point.

Proof. To show that F is non-empty, construct a sequence $\{x_n\}$ by choosing $x_n \in F_n$, $n = 1, 2, \ldots$. We can show that this sequence $\{x_n\}$ is a Cauchy sequence. If $m \geq n$, we have $F_n \supseteq F_m$ and $x_m, x_n \in F_n$. Therefore, by hypothesis we have

$$d(x_n, x_m) \leq d(F_n) \to 0 \ (n \to \infty)$$

Hence $\{x_n\}$ is a Cauchy sequence in X. Since X is complete, this sequence converges to a limit x in X, i.e.

$$d(x_n, x) \to 0 \ (n \to \infty)$$

Since each F_n is closed, it follows that $x \in \overline{F}_n = F_n$ for every n. Therefore, $x \in F$, which shows that $F = \bigcap_{n=1}^{\infty} F_n$ is non-empty. Now we have to show that F contains exactly one point. For this let F contain two distinct points x and y. Then

$$d(x, y) \leq d(F) \leq d(F_n) \to 0 \ (n \to \infty)$$

i.e. $d(x, y) = 0$ or $x = y$. Therefore, F contains exactly one point.

Remark. The converse of the above theorem is also true. Let $\{x_n\}$ be a Cauchy sequence in X and $A_n = \{x_n, x_{n+1}, \ldots\}$. Then, since $\{x_n\}$ is a Cauchy sequence, we have $d(A_n) \to 0$ as $n \to \infty$. Also $d(\overline{A}_n) \to 0$, since $d(A_n) = d(\overline{A}_n)$. Further $A_n \supset A_{n+1}$ implies $\overline{A}_n \supset \overline{A}_{n+1}$. Thus $\{A_n\}$ is a nested sequence of non-empty closed sets in X with $d(A_n) \to 0$. By our hypothesis, there exists x such that $x \in \bigcap_{n=1}^{\infty} \overline{A}_n$. Since $d(x_n, x) \leq d(\overline{A}_n) \to 0$, $x_n \to x$ in X, so that X is complete.

Remark. In the above theorem both the conditions viz. $d(F_n) \to 0$ $(n \to \infty)$ and F_n are closed sets, are necessary for the validity of the theorem. The following examples show that neither of the condition can be dropped.

Example 10. (i) Let $X = \mathbb{R}$ with the usual metric and let $F_n = [n, \infty)$. Note that $\{F_n\}$ is a sequence of non-empty closed sets such that

$$\bigcap_{1}^{\infty} F_n = \emptyset \text{ and also } d(F_n) \nrightarrow 0 \ (n \to \infty)$$

(ii) Let X be the same as in the previous example and let $F_n = \left(0, \frac{1}{n}\right]$.

Then, $d(F_n) \to 0 (n \to \infty)$. But $\bigcap_{1}^{\infty} F_n = \emptyset$, where $\{F_n\}$ is a nested sequence of non-empty sets which are not closed.

3.4 Dense and Nowhere Dense Sets

Definition 8. A subset A of a metric space (X, d) is said to be *dense* (or everywhere dense) in X iff $\overline{A} = X$

A is said to be *nowhere dense* in X if $(\overline{A})^{\circ} = \emptyset$.

A is said to be *somewhere dense* in X, if it is not nowhere dense in X.

A is said to be *dense-in-itself* if every point of A is a limit point of A, i.e. $A \subseteq A'$.

A is *perfect* if it is closed and dense-in-itself, i.e. $A = A'$. Every closed interval, the empty set $\varnothing$, and the whole space X are perfect sets.

A metric space (X, d) is said to be *separable* if it contains a countable subset which is dense in X.

Example 11. (i) $\mathbb{Q}$ is dense in $\mathbb{R}$.

Since $\mathbb{Q} \subset \mathbb{R}$, so $\overline{\mathbb{Q}} \subset \overline{\mathbb{R}} = \mathbb{R}$. We have to show that $\mathbb{R} \subset \overline{\mathbb{Q}}$. Let $x \in \mathbb{R}$. Then for every $\varepsilon > 0$ there exists $q \in \mathbb{Q}$ such that $|x - q| < \varepsilon$. Hence $x \in \mathbb{Q}$ or $x \in \mathbb{Q}'$, i.e. $x \in \mathbb{Q} \cup \mathbb{Q}' = \overline{\mathbb{Q}}$. Therefore, $\mathbb{R} \subset \overline{\mathbb{Q}}$, so that $\overline{\mathbb{Q}} = \mathbb{R}$.

(ii) Let d_∞ be the sup-metric on $C[0, 1]$. Let $P[0, T]$ be the set of all polynomials, $p(t) = a_0 + a_1 t + \ldots + a_n t^n$, with real coefficients, on $[0, T]$, $n = 0, 1, 2, \ldots$. Then $P[0, T]$ is dense in $(C[0, T], d)$.

It follows from the famous Weierstrass Approximation Theorem.

(iii) Any singleton set, any finite set, $\mathbb{N}$ and $\mathbb{Z}$ all are nowhere dense in $\mathbb{R}$. Let us consider $\mathbb{Z}$. Since $\mathbb{Z}$ is closed ($\mathbb{Z}^c$ is a union of open intervals) and so $\overline{\mathbb{Z}} = \mathbb{Z}$, But it is clear the $\mathbb{Z}$ contains no nonempty open interval.

(iv) Let $\mathbb{R}$ be the usual metric space and $A = \left\{ 1, \dfrac{1}{2}, \dfrac{1}{3}, \ldots \right\}$. Then A is nowhere dense in $\mathbb{R}$. Since $\overline{A} = \left\{ 0, 1, \dfrac{1}{2}, \dfrac{1}{3}, \ldots \right\}$ contains no nonempty open interval.

(v) The usual metric space $\mathbb{R}$ is separable.

Since $\mathbb{Q}$ is countable subset of $\mathbb{R}$ and also $\mathbb{Q}$ is dense in $\mathbb{R}$. Hence, $\mathbb{R}$ is separable.

(vi) Let $X = [0, 1]$ with the discrete metric. Then X is not separable. Since the only dense subset of the discrete metric space X is X itself.

(vii) The set $\mathbb{Q}^*$ of irrationals with the usual metric, is a separable metric space.

Let $E = \{ r + \sqrt{2} : r \in \mathbb{Q} \}$. This set is countable. Now any sphere centred at an irrational number contains a point of E. Therefore, E is dense in $\mathbb{Q}^*$ and so $\mathbb{Q}^*$ is separable.

The following result follows from the definition of separable metric spaces.

Theorem 7. A metric space (X, d) is separable iff there is a countable set $\{x_n\}$ with the following property. For each $\varepsilon > 0$ and each x in (X, d) there is at least one x_n with $d(x_n, x) < \varepsilon$.

Example 12. The space l_p $(1 \le p < \infty)$ with metric $d_p(x, y) = d(x, y)$ $= \left(\sum_{i=1}^{\infty} |x_i - y_i|^p \right)^{\frac{1}{p}}$ is separable.

Let A be the subset of l_p made up of all sequences $\{\gamma_n\}$ such that:
(i) have rational entries only, i.e. $\gamma_i \in \mathbb{Q}$, $1 \le i \le n$, $n \in \mathbb{N}$ and
(ii) have only a finite number of nonzero entries, i.e. $\gamma_i = 0$ for $i > n$ and $\gamma_n \ne 0$.

Then clearly A is countable subset of l_p. In order to show that $\overline{A} = l_p$, let $x = \{x_i\}$ be an arbitrary point in l_p. Then for a given $\varepsilon > 0$, there is an integer $N > 0$ such that

$$\sum_{i=N+1}^{\infty} |x_i|^p < \frac{\varepsilon^p}{2}$$

Since $\mathbb{Q}$ is dense in $\mathbb{R}$, there exists rational numbers $\gamma_1, \gamma_2, \ldots, \gamma_N$ such that

$$\sum_{i=1}^{N} |x_i - \gamma_i|^p < \frac{\varepsilon^p}{2}$$

Then $\gamma = \{\gamma_1, \gamma_2, \ldots, \gamma_N, 0, 0, \ldots\}$ is a point in A, and

$$[d(x, y)]^p = \sum_{i=1}^{N} |x_i - \gamma_i|^p + \sum_{i=N+1}^{\infty} |x_i|^p < \varepsilon^p$$

i.e. $d(x, y) < \varepsilon$. Hence by Theorem 7, l_p is separable.

Theorem 8. Let (X, d) be a metric space and $A \subset X$. The fol'owing are equivalent:

(i) $\overline{A}$ has empty interior
(ii) A is nowhere dense in X
(iii) $X - \overline{A}$ is dense in X

Proof. (i) $\Rightarrow$ (ii). Suppose that $(\overline{A})° = \varnothing$. Then $\overline{A}$ contains no neighbourhood, for if it did, $\overline{A} \supset S_\varepsilon(a)$ say, then $a \in (\overline{A})°$. But $a \in (\overline{A})°$ contradicts $(\overline{A})° = \varnothing$.

(ii) $\Rightarrow$ (iii). Since $\overline{X - A} = X - A°$, we get

$$(\overline{A})° = X - \overline{X - \overline{A}} \tag{1}$$

by replacing A·by $\overline{A}$ in $A° = X - \overline{X - A}$.

Since A is nowhere dense in X, i.e. $(A)° = \varnothing$, by (1) we have

$$X = \overline{X - \overline{A}}, \text{ i.e. } X - \overline{A} \text{ is dense in } X$$

(iii) $\Rightarrow$ (i). If follows easily by (1). Since $X - \overline{A}$ is dense in X, we

have
$$X = \overline{X - \overline{A}}.$$

Hence by (1), $(\overline{A})^\circ = \varnothing$, i.e. $\overline{A}$ has empty interior.

Example 13. The space c_0 is separable.

The metric on c_0 is defined by $\sup_n |x_n - y_n|$, $x, y \in c_0$. Let D be a subset of c_0 consisting of all sequences of rationals in which only a finite number of elements are different from zero. Clearly D is countable subset of c_0. To prove that D is dense in c_0, let x be any point of c_0. Now we have to find a point of D whose distance from x is arbitrarily small, i.e. we want to select a point $\gamma = \{\gamma_1, \gamma_2, \ldots,$ $\gamma_k, 0, 0, \ldots\}$, where γ_k are rationals and $\sup_k |x_k - \gamma_k|$ is small. For a given $\varepsilon > 0$, we can find n_0 such that $|x_n| < \varepsilon$ for all $n \geq n_0$. Since $\mathbb{Q}$ is dense in $\mathbb{R}$, there exists rationals $\gamma_1, \gamma_2, \ldots, \gamma_{n_0}$ such that $|x_n - \gamma_n| < \varepsilon$, $n = 1, 2, \ldots, n_0$. Then $\gamma = \{\gamma_1, \gamma_2, \ldots, \gamma_{n_0}, 0, 0, \ldots\}$ is a point in D, and $\sup_k |x_k - y_k| < \varepsilon$, i.e. $d(x, y) < \varepsilon$. Hence by Theorem 7, c_0 is separable.

Theorem 9. Let (X, d) be a metric space and $A \subset X$. Then the following statements are equivalent:

(a) A is dense in X.
(b) The only closed superset of A is X.
(c) The only open set disjoint from A is $\varnothing$.
(d) A intersects every non-empty open set.
(e) A intersects every open sphere.

Theorem 10. Let (X, d) be a metric space and $A \subset X$. Then the following statements are equivalent:

(a) A is nowhere dense in X.

(b) $\overline{A}$ does not contain any non-empty open set.

(c) Every non-empty open set has a non-empty open subset disjoint from $\overline{A}$.

(d) Every non-empty open set contains a non-empty open subset disjoint from A.

(e) Every non-empty open set contains an open sphere disjoint from A.

3.5   Categories and Baire's Category Theorem

Definition 9. Let (X, d) be a metric space and $S \subset X$. Then S is said to be of *first category* if it can be expressed as a countable union of nowhere dense sets. S is of the *second category* if it is not of the first category.

Example 14. (i) $\mathbb{R}$ is of the second category, (see Baire's category theorem). (ii) $\mathbb{Q}$ is of the first category in $\mathbb{R}$.

For $\mathbb{Q}$ is the countable union of its elements $q_1, q_2, \ldots$ and each set $\{q_i\}$, $i = 1, 2, 3, \ldots$ is obviously nowhere dense in $\mathbb{R}$.

Example 15. Any non-empty open interval of $\mathbb{R}$ is of the second category. For suppose an interval of $\mathbb{R}$ is of first category, we can write $\mathbb{R}$ as the countable union of the open intervals, that is, $\mathbb{R} = \bigcup_1^{\infty}(-n, n) = \bigcup_1^{\infty} I_n$, say. Since each I_n is of first category, $\bigcup_1^{\infty} I_n$ is of first category (note that countable union of sets of first category is of first category). Hence, $\mathbb{R}$ is of the first category, i.e. a contradiction, so each interval of $\mathbb{R}$ is of the second category.

Theorem 11 (Baire's Category Theorem). Every complete metric space is of second category.

Proof. Let $\{A_n\}$ be a sequence of nowhere dense sets in a complete metric space (X, d). Then we have to show that there exists a point in X which is not in any of the A_n's, or in otherwords, $X \neq \bigcup_1^{\infty} A_n$.

Since A_1 is nowhere dense and X is open, there exists an open sphere $S_{\varepsilon_1} \subset X$, of radius $\varepsilon_1 < 1$, such that $S_{\varepsilon_1} \cap A_1 = \varnothing$ (by Theorem 10). Let F_1 be the concentric closed sphere with radius $\frac{1}{2}\varepsilon_1$. Since A_2 is nowhere dense and F_1° is open, again it follows that F_1° contains an open sphere S_{ε_2} of radius $\varepsilon_2 < \frac{1}{2}$ such that $S_{\varepsilon_2} \cap A_2 = \varnothing$.

Let F_2 be the concentric closed sphere of radius $\frac{1}{2}\varepsilon_2$. Similarly, we can find an open sphere S_{ε_3} of radius $\varepsilon_3 < \frac{1}{2^2}$ such that $S_{\varepsilon_3} \cap A_3 = \varnothing$ and the concentric closed sphere F_3 whose radius is $\frac{1}{2}\varepsilon_3$.

Continuing this process, we get a nested sequence $\{F_n\}$ of non-empty closed subsets of X, where F_n is concentric closed sphere of radius less than $\frac{1}{2^n}$. Therefore

$$d(F_n) < \frac{2}{2^n} = \frac{1}{2^{n-1}} \to 0 \text{ as } n \to \infty$$

By the Cantor's intersection Theorem, there is a point $x \in X$ such that $\bigcup_1^{\infty} F_n = \{x\}$. Since for each n, $x \in F_n \subset S_{\varepsilon_n}$ and $S_{\varepsilon_n} \cap A_n = \varnothing$, it follows that $x \notin A_n$ for any n.

Hence $\overset{\infty}{\underset{1}{\cup}} A_n \neq X$, i.e. X is not of first category, Thus, it is of second category.

3.6   Completion of Metric Spaces.

Suppose that (X, d) is a complete metric space and that (Y, d) is an arbitrary subspace of (X, d). As given in Theorem 3, (Y, d) is complete iff Y is a closed set in (X, d). In any event, the closure $\overline{Y}$ is closed set in (X, d) and $(\overline{Y}, d)$ is complete. Moreover, Y is dense in $(\overline{Y}, d)$. Thus, in going from (Y, d) to $(\overline{Y}, d)$ we fill in any 'holes' that may exist in (Y, d). For example, let (X, d) be the set of real numbers with the usual metric, and let $Y: = \{r : r$ is rational and $0 < r < 1\}$. Then (Y, d) is not complete. However, the closure $\overline{Y} = [0, 1]$, is complete, and Y is dense in $(\overline{Y}, d)$.

Obviously the foregoing is a way to 'complete' a meric space that is a subspace of a complete metric space. Many times, however, we would like to 'complete' a metric space that is not specified as being a subspace of some complete metric space. A general application concept of completion is given as follows:

Let (X, d) be incomplete. We may use the following limits to prove that there is a complete metric space (Y, ρ) in which X is isometrically embedded as a dense subset.

Throughout $x = (x_n)$, $x' = (x'_n)$ denote sequences in X.

(i) x, x' Cauchy in (X, d) imply $(d(x_n, x'_n))$ converges in $\mathbb{R}$.

(ii) Define $\rho_1(x, x') = \lim_n d(x_n, x'_n)$ on $\mathscr{C}$ the set of all Cauchy sequences in X. Then $(\mathscr{C}, \rho_1)$ is a semimetric space.

(iii) Make $(\mathscr{C}, \rho_1)$ into a metric space (Y, ρ) by using Theorem 2, Chapter 2.

(iv) Let $Y_0 \subset Y$ be defined by using that $E \in Y_0$ if E contains a constant sequence $x_{\mathscr{C}} = \{x_1, x_2, \ldots\}$. Then Y_0 is isometric to X, i.e. $\rho(E, E') = d(x_1, x'_1)$, were E contains $x_{\mathscr{C}}$ and E' conatins $x'_{\mathscr{C}}$.

(v) Show that $\overline{Y}_0 = Y$, i.e. Y_0 is dense in Y.

(vi) Use (v) to show that (Y, ρ) is complete. The space (Y, ρ) of equivalence classes of Cauchy sequences from X, with its metric ρ, is called the completion of (X, d).

We generally define "completion" as follows:

Definition 10. Let (X, d) be a metric space. A metric space (Y, ρ) is said to be a completion of (X, d) if

(1) (Y, ρ) is complete, and

(2) (X, d) is isometric with a dense subspace (Y_0, ρ) of (Y, ρ).

Two questions can be posed. (1) Which metric space have a

completion? (2) How many "different" completions does a given space have? The answers are:
(1) Every metric space (X, d) has a completion.
(2) All the completions of (X, d) are isometric with one another, i.e. the completion is essentially unique.
We enlist the following theorems without proof.

Theorem 12. Let (X, d) be a metric space. Then (X, d) has a completion. Moreover, if (Y_1, d_1) and (Y_2, d_2) are two completions of (X, d), then (Y_1, d_1) and (Y_2, d_2) are isometric.

Theorem 13. The completion of a metric space (X, d) is separable iff (X, d) is separable.

3.7 Equivalence of Metrics

It is possible to have more than one metric on a given set and among these metrics some of them may have the same nature in relation to convergence of sequences and continuity of functions. Such metrics are known as the equivalent metrics.

Definition 11. Let d_1 and d_2 be two metrics on the set X such that

$$x_n \to x \ (n \to \infty) \text{ in } (X, d_1)$$

iff $x_n \to x \ (n \to \infty)$ in (X, d_2).
Then the two metrics d_1 and d_2 are said to be *equivalent metrics*.
We have the following result to have the equivalence of two metrics.

Theorem 14. If d_1 and d_2 are two metrics on a set X and if there exists a constant $k > 0$ such that

$$\frac{1}{k} d_1(x, y) \le d_2(x, y) \le k d_1(x, y)$$

for all $x, y \in X$, then d_1 and d_2 are equivalent.

Proof. Let $x_n \to x$ in (X, d_2). Then

$$\frac{1}{k} d_1(x_n, x) \le d_2(x_n, x)$$

implies $$x_n \to x \text{ in } (X, d_1)$$

Conversely, if $x_n \to x$ in (X, d_1), then

$$d_2(x_n, x) \le k d_1(x_n, x)$$

$$\Rightarrow \qquad x_n \to x \text{ in } (X, d_2)$$

Hence d_1 and d_2 are equivalent.

Example 16. Let d and ρ be two metrics on X, where

$$\rho(x,y) = \frac{d(x,y)}{1 + d(x,y)}$$

Then d and ρ are equivalent.

Let $x_n \to x$ in d, i.e. $d(x_n, x) \to 0$ as $n \to \infty$. Then, obviously $\rho(x_n, x) \to 0$ as $n \to \infty$ so that $x_n \to x$ in the metric ρ.

Conversely, let $x_n \to x$ in ρ, i.e. $\rho(x_n, x) \to 0$ $(n \to \infty)$. Then

$$\rho(x_n, x) = \frac{d(x_n, x)}{1 + d(x_n, x)} \geq d(x_n, x) \text{ since } d(x_n, x) + 1 > 1.$$

Hence
$$d(x_n, x) \to 0 \ (n \to \infty)$$

Therefore, two metrics d and ρ are equivalent.

3.8 Topological and Metrizable Spaces

Many of the concepts that we have defined earlier depend on the properties of open sets, e.g. closed sets, interior, closure, dense set etc. An open set itself was defined in terms of open sphere which depends strongly on the metric d. This section define the spaces in which the notion of an open set is fundamental and other notions are defined in terms of it. Such spaces are called *topological spaces*. A metric space is a special kind of topological space.

Definition 12. A topological space (X, T) is a non-empty set X of points together with a family T of subsets satisfying the following axioms:

(T_1) $X \in T$ and $\emptyset \in T$
(T_2) Any union (countable or not) of sets in T is in T.
(T_3) The interesection of any finite number of sets in T is in T.

The sets of T are called *open sets* and the family T is called *topology* for the set X.

If T is the family of all subsets of X, including $\emptyset$, then this topology is called the *discrete topology* for X ($\neq \emptyset$).

If $T = \{\emptyset, X\}$, then T is called *indiscrete topology* (or *trivial topology*) for X ($\neq \emptyset$).

Definition 13. Let S and T be two topologies for the same set X. Then S is said to be *stronger* (finer) than T if $S \supset T$. In this case we also say that T is *weaker* (coarser) than S. Thus, S is stronger than T if and only the identity map of (X, S) into (X, T) is continuous. The indiscrete topology for a set X is the weakest possible topology on X, while the discrete topology is the strongest possible topology.

Definition 14. A topological space (X, T) is called the *Hausdorff space* if for any distinct x, y in X, there exist two disjoint open sets, one containing x and the other containing y.

Definition 15. A topological space (X, T) is called *metrizable* if there exists a metric d such that $\mathcal{G} = T$, where $\mathcal{G}$ is the class of open sets determined by d.

Example 17. (i) Let X be a three-element set $X = \{a, b, c\}$ and let $T = \{\varnothing, X, \{a\}, \{b\}, \{a, b\}\}$. It is easily checked that (X, T) is a topological space.

  (ii) Let $X = \mathbb{R}$ and T be the set of all open intervals and $\varnothing$. Then (X, T) is not a topological space, as (T_2) is not satisfied.
  (iii) Let $X = [0, 1)$ in $\mathbb{R}$. Let T consist of $\varnothing$ and all sets of the form $[0, a)$, $0 < a \le 1$. Then (X, T) is a topological space.
  (iv) A metric space is a Hausdorff topological space.
  (v) Let T be the discrete topology on any set X. Then (X, T) is metrizable, for the trival metric d such that $\mathcal{G} = T$.
  (vi) Suppose X has more than one point. Let $T = \{\varnothing, X\}$ be the indiscrete topology on X. Then (X, T) is not metrizable.
  (vii) All standard classical sequence spaces w, c_0, c etc. are assumed to be the topological spaces with the topology given by their respective usual metrics.

3.9 Miscellaneous Examples

1. The space l_2 is complete with the metric

$$d_2(x, y) = \left[\sum_{i=1}^{\infty} |x_i - y_i|^2 \right]^{\frac{1}{2}}$$

It is easy to show that l_2 is a metric space. We will show that it is complete. Let $\{x^{(n)}\}$ be a Cauchy sequence in l_2, where $x^{(n)} = \{x_i^{(n)}\}_{i=1}^{\infty}$. Then for $\varepsilon > 0$, there exists an integer $N > 0$ such that

$$d(x^{(n)}, x^{(m)}) < \varepsilon, \quad \forall\, m, n \ge N \tag{1}$$

$$\Rightarrow \qquad \sum_{i=1}^{\infty} |x_i^{(n)} - x_i^{(m)}|^2 < \varepsilon^2, \quad \forall\, m, n \ge N \tag{2}$$

Therefore $\qquad |x_i^{(n)} - x_i^{(m)}| < \varepsilon, \quad \forall\, m, n \ge N$

Hence, for each fix i, $\{x_i^{(n)}\}$ is a Cauchy sequence in $\mathbb{R}$. Since $\mathbb{R}$ is complete, $x_i^{(n)} \to x_i$ $(n \to \infty)$, $i = 1, 2, \ldots$ Let $x = \{x_1, x_2, \ldots\}$. Now, since $\{x^{(n)}\} \in l_2$, we obtain

$$\sum_{i=1}^{\infty} |x_i^{(n)}|^2 \leq M < \infty \quad \text{for some } n \geq N \text{ and } M > 0$$

Hence for an integer k

$$\sum_{i=1}^{k} |x_i^{(n)}|^2 \leq M \tag{3}$$

Letting $n \to \infty$ in (3), we obtain

$$\sum_{i=1}^{k} |x_i|^2 \leq M \quad \text{for} \quad k = 1, 2, \ldots, \text{ since } x_i^{(n)} \to x_i \ (n \to \infty)$$

Again letting $k \to \infty$, we get

$$\sum_{i=1}^{\infty} |x_i^{(n)}|^2 \leq M < \infty$$

Hence $x = \{x_i\} \in l_2$.

Now from (2), we get

$$\sum_{i=1}^{\infty} (x_i^{(n)} - x_i^{(m)})^2 < \varepsilon^2, \quad m, n \geq N \tag{4}$$

Since $x_i^{(m)} \to x_i \ (m \to \infty)$, we obtain from (4)

$$\sum_{i=1}^{\infty} (x_i^{(n)} - x_i)^2 < \varepsilon^2, \quad n \geq N$$

Hence

$$\left[\sum_{i=1}^{\infty} |x_i^{(n)} - x_i|^2 \right]^{\frac{1}{2}} < \varepsilon, \quad n \geq N$$

i.e. $d(x^{(n)}, x) < \varepsilon$ for $n \geq N$. Thus $\{x^{(n)}\}$ converges to x in the metric of l_2 and $x \in l_2$. Hence l_2 is complete.

2. Homeomorphic metric spaces need not be isometric in general.

Let $X = \{1, 1/2, 1/3, \ldots\}$ and $Y = \mathbb{N}$. Let d and ρ be usual metrics on $\mathbb{R}$ restricted to X and Y. In Example 9, the function $f: X \to Y$ defined by $f(1/x) = x$ is a homeomorphism from X to Y. Now $d(x, y) = |1/x - 1/y|$ but $\rho(f(x), f(y)) = |x - y|$. Hence, $d(x, y) \neq \rho(f(x), f(y))$ for any $x, y \in X$, i.e. f is not an isometry. Therefore (X, d) and (Y, ρ) are not isometric while they are homeomorphic.

3. The space c of all convergent real sequences is complete with the metric defined by

$$d(x, y) = \sup_n |x_n - y_n|, \quad \forall \, x = (x_n), \, y = (y_n) \in c$$

Let $(x^{(n)})$ be a Cauchy sequence in c. Then

$$d(x^{(n)}, x^{(m)}) \to 0 \quad (m, n \to \infty)$$

where each member of the sequence $(x^{(n)})$ is itself a sequence,

$$x^{(n)} = (x_i^{(n)}) = (\dot{x}_1^{(n)}, x_2^{(n)}, \ldots) \in c \quad \text{for each } n$$

Now for each $\varepsilon > 0$ there exists N such that $d(x^{(n)}, x^{(m)}) < \varepsilon$, $\forall\, n, m \geq N$. Therefore,

$$|x_i^{(n)} - x_i^{(m)}| < \varepsilon \quad \text{for} \quad i = 1, 2, \ldots \text{ and } n, m \geq N \qquad (1)$$

Hence for each i, the sequence $(x_i^{(m)}) = (x_i^{(1)}, x_i^2, \ldots)$ is a Cauchy sequence in $\mathbb{R}$. Since $\mathbb{R}$ is complete, $(x_i^{(m)})$ converges to x_i say, i.e. there exists $\lim\limits_{m} x_i^{(m)} = x_i$ for each i.

Now fix $n \geq N$ and let $m \to \infty$ in (1) to get

$$|x_i^{(n)} - x_i| \leq \varepsilon \quad \text{for each } i \qquad (2)$$

Since ε is independent of i, we have

$$\sup_i |x_i^{(n)} - x_i| \leq \varepsilon, \quad \forall\, n \geq N \qquad (3)$$

and so $d(x^{(n)}, x) \to 0$ $(n \to \infty)$, where $x = (x_i)$.

Now we have to show that $x \in c$. The sequence $(x_i^{(n)}) \in c$ and so it is a Cauchy sequence. Therefore, for $\varepsilon > 0$ there is an $M = M(\varepsilon)$ such that

$$|x_i^{(N)} - x_j^{(N)}| < \varepsilon, \quad \forall\, i, j \geq M \qquad (4)$$

By (3) and (4) for $i, j \geq M$, we have

$$|x_i - x_j| = |x_i - x_i^{(N)} + x_i^{(N)} - x_j^{(N)} + x_j^{(N)} - x_j|$$

$$\leq d(x, x^{(N)}) + |x_i^{(N)} - x_j^{(N)}| + d(x^{(N)}, x) < 3\varepsilon$$

Therefore, x is a Cauchy sequence in $\mathbb{R}$. Hence $x \in c$.

4. c is a closed subspace of l_∞.

We know that $c \subset l_\infty$, so we have to show that c is closed. Let x be a limit point of c. Then there is a sequence $\{x_n\}$ in c which converges to x, i.e. $x_n \to x$ in c. Thus $\{x_n\}$ is a Cauchy sequence in c. Since c is complete, $x \in c$. Hence c is closed subspace of l_∞.

5. If E is nowhere dense in $\mathbb{R}$, then any subset of E is also nowhere dense in $\mathbb{R}$.

Let $A \subset E$. Then A does not contain any non-empty open interval, for if it contains non-empty open interval, it is contained in E which implies E is not nowhere dense in $\mathbb{R}$. Hence, a contradiction to the hypothesis. Therefore, A does not contain any nonempty open interval, and A is nowhere dense in $\mathbb{R}$.

6. If A and B are sets of first category, then $A \cup B$ is also of first category.

Let $A = \overset{\infty}{\underset{n=1}{\cup}} A_n$ and $B = \overset{\infty}{\underset{n=1}{\cup}} B_n$, where A_n and B_n are nowhere dense. Then $A \cup B$ is the finite union of countable nowhere dense sets. Hence $A \cup B$ is of first category.

7. $\mathbb{R}$ is of second category.

Suppose $\mathbb{R}$ is of first category. Then $\mathbb{R} = \overset{\infty}{\underset{n=1}{\cup}} A_n$, where each A_n is nowhere dense. We may use that each A_n is closed. Otherwise, we consider $\overline{A}_n$ in place of A_n since

$$\mathbb{R} = \overline{\mathbb{R}} = \overline{\overset{\infty}{\underset{n=1}{\cup}} A_n} = \overset{\infty}{\underset{n=1}{\cup}} \overline{A}_n$$

where each $\overline{A}_n$ is closed and nowhere dense. Let $x_1 \notin A_1$. Since A_1 is closed, there is an open interval I_1 about x_1 which does not intersect A_1. Let J_1 be a closed interval with $0 < l(J_1) < 1$ such that $J_1 \subset I_1$, where $l(J_1)$ denotes the length of J_1. Then $J_1 \cap A_1 = \varnothing$.

Now A_2 is nowhere dense and does not contain all the interior points of J_1. Take any x_2 in the interior of J_1 such that $x_2 \notin A_2$. Then there is an open interval I_2 about x_2 which does not intersect A_2 such that $I_2 \subset J_1$. Let J_2 be a closed interval with $0 < l(J_2) < 1/2$ such that $J_2 \subset I_2$. Then $J_2 \cap A_2 = \varnothing$.

Continuing this process, we obtain a sequence of non-empty closed intervals $J_1 \supset J_2 \supset J_3 \supset \ldots$ such that $0 < l(J_n) < 1/n$ and $J_n \cap A_n = \varnothing$ for every n. Hence, by the Cantor's intersection theorem, there is a point $y \in \mathbb{R}$ contained in $\overset{\infty}{\underset{n=1}{\cap}} J_n$. But for each n, y is in J_n and hence $y \notin A_n$. Hence $y \notin \overset{\infty}{\underset{n=1}{\cup}} A_n$ which is a contradiction of $\mathbb{R} = \overset{\infty}{\underset{n=1}{\cup}} A_n$, and each point $y \in \mathbb{R}$ is in some set A_n. Therefore, $\mathbb{R}$ is not of first category and is of second category.

8. A set which is nowhere dense does not mean that it is everywhere dense.

Let $X = \mathbb{R}$ with the usual metric and $A = \{1, 2, 3, \ldots\}$. Then $\overline{A} = \{1, 2, 3, \ldots\}$ which contains no non-empty open interval. Hence A is nowhere dense. But $\overline{A} \neq \mathbb{R}$. Hence, A is not dense in $\mathbb{R}$.

9. A set dense in $\mathbb{R}$ whose complement is also dense in $\mathbb{R}$.

Consider the set $\mathbb{Q}$. Every open sphere $S_r(x) = (x - r, x + r)$ contains both rational and irrational numbers. Thus, every open sphere has non-empty intersection with $\mathbb{Q}$ and $\mathbb{Q}^*$ (the set of irrationals). Hence, both $\mathbb{Q}$ and $\mathbb{Q}^*$ are dense in $\mathbb{R}$.

10. The space l_∞ of all bounded sequences is not separable.

Let E be the set of all sequences formed by 0's and 1's. E is uncountable set. Since l_∞ is a metric space with the sup-metric, we have $d(x, y) = 1$ for $x, y \in E$.

If possible, let X be an everywhere dense subset of l_∞. We shall show that X is uncountable. For this, establish a one-to-one correspondence between E and X. Let $p \in E$. Since X is everywhere dense, there is a $q \in X$ such that $d(p, q) < 1/2$. Now for any other point $s \in E$

$$1 = d(p, s) < d(p, q) + d(q, s) < 1/2 + d(q, s)$$

Therefore, we have $d(q, s) > 1/2$, i.e, we have associated a different point of X with each point of E so that X has atleast as many points of E. Hence, X cannot be a countable dense set of l_∞. Therefore, l_∞ is not separable.

11. The Cantor set is a perfect set.

The Cantor set is the set obtained from the closed interval $[0, 1]$ by removing the middle third $(1/3, 2/3)$ from $[0, 1]$, then removing the middle thirds $(1/9, 2/9)$ and $(7/9, 8/9)$ of the remaining intervals, and so on. Thus the Cantor set is the intersection of the family of sets $\{F_n : n \in \mathbb{N}\}$, where

$$F_1 = [0, 1]$$

$$F_2 = [0, 1/3] \cup [2/3, 1]$$

$$F_3 = [0, 1/9] \cup [2/9, 1/9] \cup [2/3, 7/9] \cup [8/9, 1], \text{ etc.}$$

are closed sets (since each F_n is the complement of the union of removed open intervals and the intervals $(-\infty, 0)$ and $(1, \infty)$). Hence, the Cantor set $F = \overset{\infty}{\underset{n=1}{\cap}} F_n$ is closed.

Now we show that it is dense-in-itself. Let $x \in F$, then $x = \sum_{k=1}^{\infty} \dfrac{a_k}{3^k}$, where each a_k is either 0 or 2 be the ternary expansion of x. We shall show that x is a limit point of F.

Choose the sequence (x_n) in F, such that

$$x_1 = \cdot\, a_1' a_2 a_3 \ldots a_n \ldots$$

$$x_2 = \cdot\, a_1 a_2' a_3 \ldots a_n \ldots$$

$$\vdots$$

$$x_n = \cdot\, a_1 a_2 a_3 \ldots a_n' a_{n+1} \ldots$$

where $a_n' = 0$, if $a_n = 2$ and $a_n' = 2$, if $a_n = 0$.

The sequence (x_n) of distinct points of F differ from x at the nth place in the ternary expansion. Therefore, $\lim_{n\to\infty} x_n = x$. Thus, every point of the Cantor set is a limit point of the set and so it is dense-in-itself.

12. An example of a sequence $A_1, A_2, \ldots$ of non-empty closed subsets of $\mathbb{R}$ such that

(i) $A_1 \supset A_2 \supset \ldots$,

(ii) $\bigcap_{n=1}^{\infty} A_n = \emptyset$, hold.

Consider a sequence of sets such as

$$A_1 = [1, \infty), A_2 = [2, \infty), \ldots, A_k = [k, \infty)$$

Since $[k, \infty) = (-\infty, k)^c$, $[k, \infty)$ is the complement of the open set $(-\infty, k)$ in the usual topology of $\mathbb{R}$. Hence, the sets $A_k = [k, \infty)$ are the closed sets for $k = 1, 2, \ldots$

Also $\qquad A_1 \supset A_2 \supset \ldots$ and $\bigcap_{n=1}^{\infty} A_n = \emptyset$.

EXERCISES

1. Prove that $\mathbb{R}^2$ is complete with the metric

$$d(x, y) = \sqrt{(x_1 - y_1)^2 + (x_2 - y_2)^2},$$

$$x = (x_1, x_2), y = (y_1, y_2) \in \mathbb{R}^2$$

2. Prove that $\mathbb{R}^n$ is complete with the metric

$$d(x, y) = \left[\sum_{i=1}^{n} (x_i - y_i)^2 \right]^{1/2}, \quad x = (x_1, x_2, \ldots, x_n),$$

$$y = (y_1, y_2, \ldots, y_n) \in \mathbb{R}^n$$

3. Prove that l_∞ is complete with sup-metric

$$d(x, y) = \sup_n |x_n - y_n|, x = (x_n), y = (y_n) \in l_\infty$$

4. Prove that l_p $(1 \le p < \infty)$ is complete with

$$d_p(x, y) = \left(\sum_{i=1}^{\infty} |x_i - y_i|^p \right)^{1/p}, x = (x_n), y = (y_n) \in l_p$$

5. Show that $l_\infty(p)$ and $c_0(p)$ are complete with

$$d(x, y) = \sup_n |x_k - y_k|^{p_k/M}$$

6. Define a function d on c such that (c, d) becomes a complete semi-metric space.

7. Let (X_1, d_1) and (X_2, d_2) be complete metric spaces. Show that the space (X, d) is a complete metric space, where $X = X_1 \times X_2$ and $d(x, y) = d_1(x_1, y_1) + d_2(x_2, y_2)$.

8. Show that the sequence space

$$bs(p) = \left\{ x = (x_n) \in w : \sup_k \left| \sum_{n=1}^{k} X_n \right|^{p_k} < \infty \right\}$$

is complete with the metric

$$d(x, y) = \sup_k \left| \sum_{n=1}^{k} (x_n - y_n) \right|^{p_k}, \quad \forall\, x = (x_k),\, y = (y_k) \in bs(p)$$

9. Show that the product of two separable metric spaces is a separable metric space.

10. Let $X_1 = \{1, 2, 3, \ldots\}$, $X_2 = \{1, 1/2, 1/3, \ldots\}$ and both have the usual metric. Prove that

$f: X_1 \to X_2$ defined by $f(n) = 1/n$ is a homeomorphism.

11. Let $X_1 = [0, 1]$, $X_2 = [0, 2]$ with the usual metric. Prove that $f: [0, 1] \to [0, 2]$ defined by $f(x) = ax + b$ is isometric.

12. Let $f: X_1 \to X_2$ be an isometry. Then prove that

  (i) If $\{x_n\}$ is a Cauchy sequence in X_1, then $\{f(x_n)\}$ is a Cauchy sequence in X_2.
  (ii) If $x_n \to x$ in X_1, then $f(x_n) \to f(x)$ in X_2.
  (iii) If X_1 is complete, then X_2 is complete.

13. If A is dense subset of a complete metric space X, and if $A = \bigcap_{n=1}^{\infty} G_n$ where G_n's are open in X, then prove that $X - A$ is of first category.

14. Prove that a finite union of nowhere dense sets in a metric space is a nowhere dense set.

15. Prove that the Cantor set is a closed nowhere dense subset of $\mathbb{R}$.

16. Let (X, d) be a metric space and $A \subset X$. Prove that
  (a) If X is separable then A is separable,
  (b) If Y is separable and $\bar{A} = X$, then X is separable.

17. Show that a set A in (X, d) is dense iff $B \cap A = \varnothing$ for every non-empty set B.

18. Let X be the space of all complex-valued functions $f(t)$, $-\infty < t < \infty$, such that

$$\lim_{T \to \infty} \frac{1}{2T} \int_{-T}^{T} |f(t)|^2 \, dt < \infty$$

Let the metric on X be given by

$$d(x, y) = \left[\lim_{T \to \infty} \frac{1}{2T} \int_{-T}^{T} |f(t) - g(t)|^2 \, dt \right]^{1/2}$$

Show that (X, d) is not separable.

19. Show that the discrete metric space X_d is separable iff X is countable.
20. Prove that a closed set is nowhere dense iff it contains no open set.
21. Show that the usual metric space $\mathbb{C}$ is separable.
22. Show that $(C[a, b], d_\infty)$ is separable.
23. Prove that if X is of second category, and if $X = A \cup B$, then either A or B must be of second category.
24. Prove that X is of second category in itself if and only if the intersection of every countable family of dense open sets in X is non-empty.
25. Prove that l_1 is dense but of the first category in c_0.
26. Prove that l_∞ is dense but of the first category in w.
27. Show that the set of irrationals is of second category.
28. Use Baire's Category Theorem to prove the existence of every where continuous, nowhere differentiable real-valued functions.

4

Continuity

The introduction of abstract metric spaces allows the generalization of the concept of continuity. This generalization is one of the main reasons for investigating metric spaces.

4.1 Limit of Function

First we introduce the notion of the limit of a function in metric spaces.

Let (X, d_1) and (Y, d_2) be metric spaces and let $a \in X$. Let f be a function whose range is contained in Y and whose domain contains all $x \in X$ such that $d_1(a, x) < r$ for some $r > 0$ except possibly at $x = a$. Also assume that a is a cluster point of the domain of f i.e. for every $r > 0$, there is a point b in the domain of f distinct from a such that $d_1(a, b) < r$.

Definition 1. The function f is said to have *limit* ℓ ($\ell \in Y$) at a ($a \in X$), if for a given $\varepsilon > 0$ there exists a $\delta > 0$ such that

$$d_2(f(x), \ell) < \varepsilon \text{ whenever } 0 < d_1(x, a) < \delta$$

Write $\qquad \lim_{x \to a} f(x) = \ell \text{ or } f(x) \to \ell \text{ as } x \to a$

Following theorem gives the characterization of the limit of a function in terms of the limit of a sequence.

Theorem 1. Let (X, d_1) and (Y, d_2) be metric spaces and $a \in X$. Let $f : X \to Y$. Then

   (i) $\lim_{x \to a} f(x) = \ell$

If and only if

   (ii) $\lim f(x_n) = \ell$

for every sequence $\{x_n\} \subset X$ such that

$$x_n \neq a, x_n \to a \text{ as } n \to \infty$$

Proof. Suppose $\lim_{x \to a} f(x) = \ell$. Let $\varepsilon > 0$ be given. Then there exists a $\delta > 0$ such that

$$d_2(f(x),\, \ell) < \varepsilon \text{ whenever } 0 < d_1(x, a) < \delta \tag{1}$$

Now choose $\{x_n\} \subset X$ such that $x_n \to a (n \to \infty)$, and $x_n \neq a$, for every $n = 1, 2, \ldots$. Therefore, there is a positive integer N such that for $n \geq N$, we have

$$0 < d_1(x_n, a) < \delta$$

Thus by (1)

$$d_2(f(x_n),\, \ell) < \varepsilon \quad \text{for all } n \geq N \tag{2}$$

Since ε being arbitrary, (2) yields

$$\lim_{n \to \infty} f(x_n) = \ell$$

Conversely, suppose (ii) holds, but $\lim_{x \to a} f(x) \neq \ell$. Therefore, there is an $\varepsilon > 0$ such that for every $\delta > 0$,

$$d_2(f(x),\, \ell) \geq \varepsilon \quad \text{when } 0 < d_1(x, a) < \delta \tag{3}$$

Take a sequence $\delta = \{\delta_n\}$, $\delta_n = 1/n$, $n = 1, 2, \ldots$ Then we have $x_n \in X$ for each n and

$$0 < d_1(x_n, a) < \frac{1}{n}$$

From (3) it follows that

$$d_2(f(x_n),\, \ell) \geq \varepsilon$$

While $x_n \to a$. This contradicts (ii). Hence, (i) must hold.

The following theorem gives the algebraic properties of the limits of real valued functions on metric spaces.

Theorem 2. Let (X, d) be a metric space and let a, be a point in X. Let f and g be real valued functions whose domains are subsets of X and ranges are in $\mathbb{R}$ with the usual absolute value metric. If $\lim_{x \to a} f(x) = L$ and $\lim_{x \to a} g(x) = M$, then

(i) $\lim_{x \to a}[f(x) \pm g(x)] = L \pm M$

(ii) $\lim_{x \to a}[f(x) \cdot g(x)] = L M$

(iii) $\lim_{x \to a} \dfrac{f(x)}{g(x)} = \dfrac{L}{M},\, M \neq 0$

Proof follows exactly on the same lines as in case of $\mathbb{R}$.

4.2 Continuous Functions

As in generalization of the limits of sequences and functions in metric spaces, we define continuous functions in a metric space (X, d) by replacing the absolute value in the definition of continuity in $\mathbb{R}$ by the metric.

Definition 2. Let $f : X \to Y$ be mapping of the metric space (X, d_1) into the metric space (Y, d_2). The mapping f is said to be *continuous* at the point x_0 in X if for every real number $\varepsilon > 0$, there exists a real number $\delta > 0$ such that $d_2(f(x), f(x_0)) < \varepsilon$ whenever $d_1(x, x_0) < \delta$. The mapping f is said to be *continuous* if it is continuous at each point in its domain. A function which is not continuous at some point is said to be *discontinuous* at that point.

Example 1. (i) Let $f: \ell_2 \to \ell_2$ be defined by

$$f(x) = \{0, x_1, x_2, \ldots\}, \quad x = \{x_k\} \in \ell_2$$

and the metric on ℓ_2 be the usual metric

$$d(x, y) = d_2(x, y) = \left(\sum_{k=1}^{\infty} |x_k - y_k|^2 \right)^{\frac{1}{2}}$$

Then f is continuous on ℓ_2.

(ii) Let $f: [0, 1] \to \mathbb{R}$ be defined by

$$f(x) = \begin{cases} x, & x \text{ is rational,} \\ 1 - x, & x \text{ is irrational.} \end{cases}$$

Then f is continuous only at $x = \dfrac{1}{2}$ in $[0, 1]$.

(iii) Let us consider the Dirichlet function $f : \mathbb{R} \to \mathbb{R}$ defined by

$$f(x) = \begin{cases} 1, & x \text{ is rational,} \\ 0, & x \text{ is irrational.} \end{cases}$$

Then f is discontinuous at every point of $\mathbb{R}$.
Definition 2 can be given in terms of open spheres as follows:

Definition 3. A function $f:(X, d_1) \to (Y, d_2)$ is said to be continuous at a point $x_0 \in X$ if for every $\varepsilon > 0$ there exists $\delta > 0$ such that

$$x \in S_\delta(x_0) \Rightarrow f(x) \in S_\varepsilon(f(x_0))$$

i.e. $\qquad\qquad f(S_\delta(x_0)) \subset S_\varepsilon(f(x_0))$

Example 2. Let $X = \mathbb{R}^n$ and $Y = \mathbb{R}^m$ and $f : X \to Y$. Then f can be represented with matrix formulation as

$$
\begin{bmatrix} y_1 \\ y_2 \\ \vdots \\ y_m \end{bmatrix} = \begin{bmatrix} a_{11} & a_{12} \cdots a_{1n} \\ a_{21} & a_{22} \cdots a_{2n} \\ \vdots & \\ a_{m1} & a_{m2} \cdots a_{mn} \end{bmatrix} \begin{bmatrix} x_1 \\ x_2 \\ \vdots \\ x_n \end{bmatrix}
$$

where a_{ij}'s are real numbers. Consider the following metrics on X and Y.

For X: $d(u, v) = \{| u_1 - v_1 |^2 + \ldots + | u_n - v_n |^2\}^{1/2}$

For Y: $d(w, z) = \{| w_1 - z_1 |^2 + \ldots + | w_m - z_m |^2\}^{1/2}$

Let x_0 be an arbitrary element in X, i.e.

$$
x_0 = (x_{01} \quad x_{02} \ldots x_{0n})^T
$$

The image of x_0 under a mapping f is given by

$$
y_0 = \begin{bmatrix} y_{01} \\ y_{02} \\ \cdot \\ \cdot \\ y_{0m} \end{bmatrix} = \begin{bmatrix} a_{11}x_{01} + a_{12}x_{02} + \ldots + a_{1n}x_{0n} \\ a_{21}x_{01} + a_{22}x_{02} + \ldots + a_{2n}x_{0n} \\ \cdot \\ \cdot \\ a_{m1}x_{01} + a_{m2}x_{02} + \ldots + a_{mn}x_{0n} \end{bmatrix}
$$

It follows that if x is another arbitrary point in X and y is its image, then

$$
d(y, y_0)^2 = \sum_{i=1}^{m} \left| \sum_{j=1}^{n} a_{ij}(x_j - x_{0j}) \right|^2
$$

Using Schwarz inequality, it follows that

$$
d(y, y_0)^2 \leq \sum_{i=1}^{m} \left(\sum_{j=1}^{n} |a_{ij}|^2 \right) \left(\sum_{j=1}^{n} |x_j - x_{0j}|^2 \right)
$$

$$
\leq M^2 d(x, x_0)^2, \quad M = (\Sigma_{ij} |a_{ij}|^2)^{1/2}
$$

Then given an $\varepsilon > 0$ choose $\delta = \varepsilon/M$, provided $M \neq 0$. It is clear that $d(y, y_0) < \varepsilon$ whenever $d(x, x_0) < \delta$. Hence, every such mapping f is continuous.

Theorem 3. (a) Let (X, d_1), (Y, d_2) and (Z, d_3) be metric spaces and let $f: X \to Y$, $g: Y \to Z$. If f is continuous at $a \in X$ and g is continuous at $f(a) \in Y$, then $g \cdot f$ is continuous at a.

(b) If f and g are continuous real valued functions on a metric space X, then $f \pm g$ and $f \cdot g$ are continuous. Further if $g(x) \neq 0$ for each x in X, then f/g is also continuous on X.

Proof. (a) Let $\{x_n\}$ be a sequence in X such that $x_n \to a$ as $n \to \infty$. To prove the theorem, we have to show that

$$\lim_{n \to \infty} g(f(x_n)) = g(f(a))$$

Since f is continuous at a, we have $\lim_{n \to \infty} f(x_n) = f(a)$.

Let $y_n = f(x_n)$ and $y_n \to f(a)$ as $n \to \infty$ in (Y, d_2). Since g is continuous, $\lim_{n \to \infty} g(y_n) = g(f(a))$. Substituting for y_n, we get

$$\lim_{n \to \infty} g(f(x_n)) = g(f(a))$$

Hence $g \cdot f: X \to Z$ is continuous.

(b) We shall only prove the continuity of $f + g$. Others can be proved in a similar manner.

Let $$x_n \to x \text{ and } y_n \to y \text{ in } (X, d)$$

Since f and g are continuous from X to $\mathbb{R}$, $f(x_n) \to f(x)$ and $g(y_n) \to g(y)$ in $\mathbb{R}$. By using the properties of limits, we have $f(x_n) + g(y_n) \to f(x) + g(x)$ in $\mathbb{R}$ with the usual metric so that

$$\lim_{n \to \infty} (f + g)(x_n) = (f + g)x$$

Hence $f + g$ is continuous.

4.3 Characterizations of Continuous Functions

Theorem 4. Let (X, d_1) and (Y, d_2) be metric spaces and f be a function of X into Y. Then f is continuous at $x_0 \in X$ iff any one and hence both of the following conditions hold:

(i) Given $\varepsilon > 0$ there exists $\delta > 0$ such that $S_\delta(x_0) \subset f^{-1}(S_\varepsilon(f(x_0)))$.

(ii) Whenever $\{x_n\}$ is a sequence of points in X such that $x_n \to x_0$, then $f(x_n) \to f(x_0)$, where $\{f(x_n)\}$ is a sequence of points in Y.

Proof. (i) Assume that f is continuous. Then from Definition 3, we obtain

$$x \in S_\delta(x_0) \Rightarrow f(x) \in S_\varepsilon(f(x_0))$$

i.e. $$x \in S_\delta(x_0) \Rightarrow x \in f^{-1}(S_\varepsilon(f(x_0)))$$

Hence $$S_\delta(x_0) \subset f^{-1}(S_\varepsilon(f(x_0)))$$

Conversely, if $S_\delta(x_0) \subset f^{-1}(S_\varepsilon(f(x_0)))$, then $f(S_\delta(x_0)) \subset S_\varepsilon(f(x_0))$. This shows that f is continuous at x_0, by Definition 3.

(ii) Suppose f is continuous at $x_0 \in X$. Let $\varepsilon > 0$ be given and $\{x_n\}$ be a sequence in X such that $x_n \to x_0$. Then there exists a $\delta > 0$ such that $f(S_\delta(x_0)) \subset S_\varepsilon(f(x_0))$. Since $x_n \to x_0$, there is a positive integer N such that $x_n \in S_\delta(x_0)$ for all $n \geq N$. Hence $f(x_n) \in S_\varepsilon(f(x_0))$ for all $n \geq N$, i.e. $f(x_n) \to f(x_0)$.

Conversely, assume that for every sequence $\{x_n\}$, $x_n \to x_0$ implies $f(x_n) \to f(x_0)$. Let, if possible, f is not continuous at x_0. Then, there is an $\varepsilon > 0$ such that there is no open sphere centred on x_0 whose f - image is contained in $S_\varepsilon(f(x_0))$. Consider a sequence of open spheres $\{S_{1/n}(x)\}_{n=1}^{\infty}$. Let $x_n \in S_{1/n}(x_0)$ be such that $f(x_n) \notin S_\varepsilon(f(x_0))$. Hence, $d_1(x_n, x_0) < 1/n$ but $d_2(f(x_n), f(x_0)) > \varepsilon$. This contradicts the fact that $f(x_n) \to f(x_0)$ as $n \to \infty$. Hence, f is continuous at x_0.

Consequently, f is continuous iff $x_n \to x \Rightarrow f(x_n) \to f(x)$ for every $\{x_n\} \subset X$.

Theorem 5. Let (X, d_1) and (Y, d_2) be metric spaces and let $f : X \to Y$. Then the following are equivalent

(i) $f : X \to Y$ is continuous
(ii) $f^{-1}(G)$ is open in X whenever G is open in Y.
(iii) $f^{-1}(F)$ is closed in X whenever F is closed in Y.

Proof. (i) $\Rightarrow$ (ii). Take $x \in f^{-1}(G)$. Then $y = f(x) \in G$. Since G is open, there is an open sphere $S_\varepsilon(y) \subset G$ for some $\varepsilon > 0$. Hence, $f^{-1}(S_\varepsilon(y))$ contains an open sphere $S_\delta(x)$ by Theorem 4. Thus we have $f^{-1}(G) \supset f^{-1}(S_\varepsilon(y)) \supset S_\delta(x)$. Therefore, $f^{-1}(G)$ is open in X.

(ii) $\Rightarrow$ (iii). Let F be closed in Y. Then $Y - F$ is open in Y and by (ii), $f^{-1}(Y - F)$ is open in X. But $f^{-1}(Y - F) = f^{-1}(Y) - f^{-1}(F) = X - f^{-1}(F)$. Hence $X - f^{-1}(F)$ is open set in X so that $f^{-1}(F)$ is closed in X.

(iii) $\Rightarrow$ (i). Let G be any open set in Y. Then $Y - G$ is closed in Y and hence by (iii), $f^{-1}(Y - G)$ is closed in X. Therefore, as above $X - f^{-1}(G)$ is closed in X and so $f^{-1}(G)$ is open in X. Let $x \in X$ be arbitrary and $\varepsilon > 0$ be given. Then, $f(x) \in Y$ and $S_\varepsilon(f(x))$ is open in Y and hence $f^{-1}(S_\varepsilon(f(x)))$ is open in X. Since $x \in f^{-1}(S_\varepsilon(f(x)))$, there is an open sphere $S_\delta(x) \subset f^{-1}(S_\varepsilon(f(x)))$. Hence, by Theorem 4, f is continuous at x. Since x was arbitrary, f is continuous.

Theorem 6. Let (X, d_1) and (Y, d_2) be metric spaces and $f : X \to Y$. Then f is continuous iff any one and hence both of the following conditions hold.

(i) $f(\overline{A}) \subset \overline{f(A)}$, for every subset A of X.

(ii) $\overline{f^{-1}(B)} \subset f^{-1}(\overline{B})$, for every subset B of Y.

Proof. (i) Let f be continuous. Then, since $\overline{f(A)}$ is closed in Y, $f^{-1}(\overline{f(A)})$ is closed in X by Theorem 5. Now, we have $f(A) \subset (\overline{f(A)}) \Rightarrow A \subset f^{-1}(\overline{f(A)}) \Rightarrow \overline{A} \subset \overline{f^{-1}(\overline{f(A)})} \Rightarrow \overline{A} \subset f^{-1}(\overline{f(A)})$, siince $f^{-1}(\overline{f(A)})$ is closed. Hence, $f(\overline{A}) \subset \overline{f(A)}$.

Conversely, let $f(\overline{A}) \subset \overline{f(A)}$, for every subset A of X. We shall prove that f is continuous. Let F be any closed subset in Y. Then $\overline{F} = F$. Now $f^{-1}(F)$ is a subset of X and so, by the hypothesis, we have $f(\overline{f^{-1}(F)}) \subset \overline{f(f^{-1}(F))} \Rightarrow f(\overline{f^{-1}(F)}) \subset \overline{F} = F \Rightarrow \overline{f^{-1}(F)} \subset f^{-1}(F)$. But $f^{-1}(F) \subset \overline{f^{-1}(F)}$. Therefore, $\overline{f^{-1}(F)} = f^{-1}(F)$, i.e. $f^{-1}(F)$ is closed in X. Thus, by Theorem 5, f is continuous.

(ii) Let f be continuous. Write $A = f^{-1}(B)$. Then $f(A) \subset B \Rightarrow \overline{f(A)} \subset \overline{B} \Rightarrow f(\overline{A}) \subset \overline{B}$, since $\overline{f(A)} \subset f(\overline{A})$ by (i). Therefore, $\overline{A} \subset f^{-1}(\overline{B})$. Hence $\overline{f^{-1}(B)} \subset f^{-1}(\overline{B})$.

Conversely, let $\overline{f^{-1}(B)} \subset f^{-1}(\overline{B})$, for every subset B of Y. Suppose that F is closed set in Y. Then $\overline{F} = F$. Now, by the hypothesis, we have

$$\overline{f^{-1}(F)} \subset f^{-1}(\overline{F}) = f^{-1}(F), \text{ since } \overline{F} = F$$

But $f^{-1}(F) \subset \overline{f^{-1}(F)}$, so that $\overline{f^{-1}(F)} = f^{-1}(F)$. Hence $f^{-1}(F)$ is closed in X and by Theorem 5, f is continuous.

Theorems 4, 5 and 6 can be combined in a single result as follows.

Theorem 7. Let (X, d_1) and (Y, d_2) be metric spaces and $f: X \to Y$ be a function. Then the following are equivalent:

(i) f is continuous

(ii) for every $\{x_n\} \subset X$ with $x_n \to x$, $f(x_n) \to f(x)$ for each $x \in X$.

(iii) $f^{-1}(G)$ is open in X whenever G is open in Y.

(iv) $f^{-1}(F)$ is closed in X whenever F is closed in Y.

(v) $f(\overline{A}) \subset \overline{f(A)}$, for every subset $A \subset X$.

(vi) $\overline{f^{-1}(B)} \subset f^{-1}(\overline{B})$, for every subset $B \subset Y$.

4.4   Some Further Results

Remark 1. Note that: (i) under a continuous map, the image of an open set need not be an open set, and (ii) if f is continuous one-to-one map, then f^{-1} need not be continuous.

(i) Let $X = \mathbb{R}$ with discrete metric and let $Y = \mathbb{R}$ with the usual metric and $F : X \to Y$ be defined by $f(x) = x$. Since X is discrete, every subset of X is open. Hence, for any open set G of $Y, f^{-1}(G)$ is open in X so that f is continuous.

Let $A = \{x\}$. Then A is open in X. But $f(A) = \{x\}$ is not open in Y.

(ii) As in (i), let X and Y be two metric spaces and let $f : X \to Y$. First note that f is continuous. But $f^{-1} : Y \to X$ is not continuous. For $\{x\}$ is an open set in X. But we have $(f^{-1})^{-1}(x) = \{x\}$ which is not open in Y. Therefore, f^{-1} is not continuous.

Theorem 8. Let G be an open subset of $\mathbb{R}$. Then the characteristic function χ_G is continuous at each point of G.

Proof. Let $x \in G$ and let $\{x_n\}$ be a sequence in G, such that $x_n \to x$ in G. Now to show that χ_G is continuous on G, we have to show $\chi_G(x_n) \to \chi_G(x)$.

Now $\{\chi_G(x_n)\} = \{1, 1, 1, \ldots\ldots\}$ so $\chi_G(x_n) \to 1$. But $\chi_G(x) = 1$.

Hence $\chi_G(x_n) \to \chi_G(x)$ so that χ_G is continuous at every point of G.

Theorem 9. Let f and g be continuous real-valued functions on the metric space X. Let A be the set of all points $x \in X$ such that $g(x) < f(x)$. Then A is open.

Proof. If f and g are continuous real-valued functions on X to $\mathbb{R}$, then $f - g$ is also a continuous real-valued function on X to $\mathbb{R}$. If $g(x) < f(x)$, then $f(x) - g(x) > 0$.

Now

$$A := \{x \in X : g(x) < f(x)\}$$

$$:= \{x \in X : f(x) - g(x) > 0\}$$

$$:= \{x \in X : f(x) - g(x) \in (0, \infty)\}$$

$$:= \{x \in X : (f - g)(x) \in (0, \infty)\}$$

$$= (f - g)^{-1}(0, \infty).$$

Hence $A = (f - g)^{-1}(0, \infty)$, since $(0, \infty)$ is open in $\mathbb{R}$, and $(f - g)$ is continuous, $(f - g)^{-1}(0, \infty)$ is open in X. Therefore, A is an open set in X.

Theorem 10. Let f be a continuous real-valued function on a metric space X. Let A be the set of all points $x \in X$ such that $f(x) \geq 0$. Then A is closed.

Proof. Let $x \in X$ be a limit point of A. We have to show that $x \in A$, i.e. $f(x) \geq 0$. Since x is a limit point of A, there exists a sequence x_n in A converging to x. Since f is continuous real-valued function on X, $f(x_n) \to f(x)$ as $n \to \infty$. Since each

$$f(x_n) \geq 0, \quad f(x) \geq 0 \text{ so that } x \in A$$

Hence, A is a closed set.

4.5 Uniform Continuity

The notion of continuity is local in character. We observe that in the definition of continuity δ, in general, may depend not only on ε but also on the point x_0. On the other hand, uniform continuity is that idea of continuity where δ does not depend on x_0 but only on ε, i.e. for each $\varepsilon > 0$ we can find $\delta > 0$ which works uniformly over the entire space X. Thus, the notion of uniform continuity is global in character.

Definition 4. A mapping f of a metric space (X, d_1) into a metric space (Y, d_2) is said to be *uniformly continuous* if given $\varepsilon > 0$, there exists a $\delta > 0$ (depending on ε only) such that

$$d_2(f(x), f(x_0)) < \varepsilon \text{ holds whenever } d_1(x, x_0) < \delta, x, x_0 \in X$$

Remark 2. Every uniformly continuous function is also continuous but converse need not be true. A function which is not continuous is also not uniformly continuous.

The following example shows that a continuous function need not be uniformly continuous.

Example 3. Let $\mathbb{R}$ be the usual metric space and $f : \mathbb{R} \to \mathbb{R}$ be defined by $f(x) = x^2$.

It is easy to verify that f is continuous.

Choose $\varepsilon = 1$. Let $\delta > 0$ be given. Take

$$x = \frac{\delta}{2} + \frac{1}{\delta} \quad \text{and} \quad y = \frac{1}{\delta}$$

Then

$$|x - y| = \frac{\delta}{2} < \delta$$

But

$$|f(x) - f(y)| = \left| \left(\frac{\delta}{2} + \frac{1}{\delta} \right)^2 - \frac{1}{\delta^2} \right| = \left| \frac{\delta^2}{4} + 1 \right| = \frac{\delta^2}{4} + 1 > 1 = \varepsilon$$

Thus, $|x - y| < \delta$ does not imply $|f(x) - f(y)| < \varepsilon$, and so f is not uniformly continuous.

Example 4. Let $X = Y = L_2[0, T] = \left\{ x(t): \int_0^T |x(t)|^2 \, dt < \infty \right\}$, where $0 < T < \infty$, be given with the usual metric

$$d_2(x, y) = \left(\int_0^T |x(t) - y(t)|^2 \, dt \right)^{\frac{1}{2}}$$

Then $y(t) = \int_0^t x(u) \, du$ represents a mapping f of $L_2[0, T]$ into itself. Moreover, f is continuous. Let x_0 and x be arbitrary points in X, and let $y_0 = f(x_0)$ and $y = f(x)$.
Then

$$|y(t) - y_0(t)| = \left| \int_0^t [x(u) - x_0(u)] \, du \right| \le \int_0^t |x(u) - x_0(u)| \, du$$

$$\le \left\{ \int_0^T 1^2 \, du \right\}^{1/2} \left\{ \int_0^T |x(u) - x_0(u)|^2 \, du \right\}^{1/2}$$

by the Schwarz Inequality

$$= \sqrt{T} \left\{ \int_0^T |x(u) - x_0(u)|^2 \, du \right\}^{\frac{1}{2}}$$

Therefore, $\int_0^T |y(t) - y_0(t)|^2 \, dt \le T^2 \int_0^T |x(u) - x_0(u)|^2 \, du$

or $\qquad\qquad d_2(y, y_0) \le T d_2(x, x_0)$

This last inequality implies that f is uniformly continuous.

Theorem 11. Let (X, d) be a metric space and $A \subset X$. Then, the function $f : X \to \mathbb{R}$ given by

$$f(x) = d(x, A), x \in X$$

is uniformly continuous.

Proof. By triangle inequality

$$d(x, a) \le d(x, y) + d(y, a), \text{ for every } a \in A \text{ and } x, y \in X$$

Then $\qquad\qquad \inf_{a \in A} d(x, a) \le d(x, y) + \inf_{a \in A} d(y, a)$

$\Rightarrow$ $\qquad d\,(x, A) \le d(x, y) + d(y, A)$

Hence $\qquad d\,(x, A) - d(y, A) \le d\,(x, y)$

This is true for all $x, y \in X$. Therefore, on interchanging x and y, we get

$$d\,(y, A) - d(x, A) \le d(x, y)$$

Thus

$$|\,d(x, A) - d(y, A)\,| \le d(x, y)$$

Therefore, for a given $\varepsilon > 0$, choosing a δ such that $0 < \delta \le \varepsilon$, we have

$$|f\,(x) - f\,(y)| = |\,d\,(x, A) - d\,(y, A)| \le d\,(x, y) < \delta \le \varepsilon$$

Hence, f is uniformly continuous on X.

Theorem 12. Let (X, d_1) and (Y, d_2) be metric spaces and $f : X \to Y$ be a uniformly continuous function. If $\{x_n\}$ is a Cauchy sequence in X, then $\{f\,(x_n)\}$ is a Cauchy sequence in Y.

Proof. Since f is uniformly continuous, for a given $\varepsilon > 0$, there exists a $\delta > 0$ such that

$$d_2(f\,(x), f\,(x')) < \varepsilon \text{ whenever } d_1(x, x') < \delta, \, x, x' \in X$$

In particular, we have

$$d_2\,(f\,(x_n), f\,(x_m)) < \varepsilon \quad \text{whenever} \quad d_1(x_n, x_m) < \delta \qquad (1)$$

But $\{x_n\}$ is a Cauchy sequence in X. Therefore, given $\delta > 0$, there exists an integer $N > 0$ such that

$$d_1(x_n, x_m) < \delta \text{ for every } n, m \ge N \qquad (2)$$

Therefore from (1) and (2) it follows that

$$d_2(f\,(x_n), f\,(x_m)) < \varepsilon \text{ for every } n, m \ge N$$

Hence $\{f\,(x_n)\}$ is a Cauchy sequence in Y.

Remark 3. Note that a continuous function may not send a Cauchy sequence to a Cauchy sequence.

For example, let $X = \mathbb{R}^+$ with the usual metric d, and $Y = \mathbb{R}$ with the usual metric d'. Then $f : X \to Y$ defined by $f\,(x) = 1/x$, $\forall\, x \in X$, is continuous on X. Now $\{1/n\}$ is a Cauchy sequence in X. But $f\,(1/n) = \{n\}$ is not Cauchy in Y.

4.6   Relation Between Continuity and Homeomorphism

We discussed the concept of homeomorphism in Chapter 3. Now we give relations of homeomorphism with a continuous open map and a continuous closed map.

Definition 5. A function $f: X \to Y$ is said to be an *open map* if $f(G)$ is open in Y for every open set G in X; and f is said to be a *closed map* if $f(F)$ is closed in Y for every closed set F in X.

Recall that the concept of homeomorphism involves the continuity of f and f^{-1}. Homeomorphism maps open (closed) subsets of X onto open (closed) subsets of Y.

Theorem 13. A one-to-one function $f: X \to Y$ is a homeomorphism iff f is a continuous open map.

Proof. If f is a homeomophism from X onto Y, it maps open sets of X onto Y. Hence f is continuous open map.

Conversely, suppose that $f: X \to Y$ is a continuous open map. For this it is enough to prove that f^{-1} is continuous from Y onto X.

Note that f^{-1} is well-defined and is continuous iff for every open set G in X, $(f^{-1})^{-1}(G)$, is open in Y.

But $(f^{-1})^{-1}(G) = f(G)$, Therefore, f^{-1} is continuous iff for every open set G in X, $f(G)$ is open in Y. Hence f^{-1} is continuous iff f is an open map. Hence, f is a homeomorphism.

Theorem 14. A one-to-one function f on X onto Y is a homeomorphism iff f is continuous closed map from X to Y.

The proof can be obtained easily as in Theorem 13 by using Theorem 5.

Theorems 13 and 14 can also be combined together.

Theorem 15. If $f: X \to Y$ is a one-to-one and onto map, then the following are equivalent:

   (i) f is continuous open map

   (ii) f is continuous closed map

   (iii) f is homeomorphism.

Theorem 16. Let f be an isometry from (X, d_1) onto (Y, d_2). Then f and f^{-1} are both continuous. Moreover, f is a homeomorphism.

Proof. Since X and Y are isometric, $f: X \to Y$ is bijective. Hence $f^{-1}: Y \to X$ is well defined.

Let $x_n \to x$ in (X, d_1). Since f is an isometry, we have $d_1(x_n, x) = d_2(f(x_n), f(x))$. Thus $x_n \to x$ implies $d_1(x_n, x) \to 0$ $(n \to \infty)$. Consequently, $d_2(f(x_n), f(x)) \to 0$ $(n \to \infty)$ in Y, so that $f(x_n) \to f(x)$ in Y. Hence, by Theorem 4, f is continuous.

Now, let $y_n \to y$ in Y. Since f is bijective, there exist distinct x_n and x in X such that $f(x_n) = y_n$ and $f(x) = y$, so that $x_n = f^{-1}(y_n)$ and $x = f^{-1}(y)$. But by isometry

$$d_1(f^{-1}(y_n), f^{-1}(y)) = d_1(x_n, x) = d_2(f(x_n), f(x)) = d_2(y_n, y)$$

Hence, $y_n \to y$ implies $d_2(y_n, y) \to 0$ and consequently

$$d_1(f^{-1}(y_n), f^{-1}(y)) \to 0 \ (n \to \infty) \text{ in } X$$

i.e. $\quad\quad f^{-1}(y_n) \to f^{-1}(y)$

Hence f^{-1} is continuous. Therefore, f is a homeomorphism.

4.7 Continuous Function on a Topological Space

Motivated by Theorem 5 we now define the continuity for functions on topological spaces.

Definition 6. Let X and Y be topological spaces. Then $f : X \to Y$ is called *continuous* on X if the inverse image of every open set in Y is open in X.

Example 7. (i) Let X have the discrete topology and let Y be any topological space. Then every function $f : X \to Y$ is necessarily continuous on X. For $f^{-1}(G)$, where G is open set in Y, is a subset of X and so open.

(ii) Let $X = \{x, y, z\}$ and $T = \{\varnothing, X, \{x\}, \{y\}, \{x, y\}\}$ so that (X, T) is a topological space. Define $f : X \to Y$ by $f(x) = x$, $f(y) = z$, $f(z) = y$. Then, by considering inverse images of the sets of T, we find that f is not continuous.

The concept of sequential continuity in a topological space is sometimes very useful.

Definition 7. If $\{x_n\}$ is a sequence in a topological space X, then we say that $\{x_n\}$ *converges* to $x \in X$ (written $x_n \to x$) if for every open set G containing x there exists $N = N(G)$ such that $n > N$ implies $x_n \in G$.

Definition 8. Let $f : X \to Y$, where X, Y are topological spaces. Then f is called *sequentially continuous* at a point $x \in X$ if for every sequence $x_n \to x$ (in X) we have $f(x_n) \to f(x)$ (in Y).

Now we can rewrite Theorem 4 as follows.

Theorem 17. (i) If X and Y are topological spaces and if $f : X \to Y$ is continuous, then f is sequentially continuous on X, but not conversely in general.

(ii) If X and Y are metric spaces, then sequential continuity on X implies continuity on X.

Proof. (i) Take any $x \in X$ and any open set G containing $f(x)$. Then $f^{-1}(G)$ contains x and is open in X. If $x_n \to x$, then $x_n \in f^{-1}(G)$ for almost all n, whence $f(x_n) \in G$ for almost all n, i.e. $f(x_n) \to f(x)$. Thus, f is sequentially continuous at each point of X.

Conversely, let T consist of $\varnothing$ and the complements of countable sets in $\mathbb{R}$. Then T is a topology for $\mathbb{R}$ and a sequence $\{x_n\}$ converges to x in this topology iff $x_n = x$ for all sufficiently large n. Also every function $f: (\mathbb{R}, T) \to (X, T')$, where (X, T') is any topological space, is sequentially continuous at each point of $\mathbb{R}$: and $f(x) = x$ is sequentially continuous but not continuous on $\mathbb{R}$.

(ii) This follows from Theorem 4.

4.8   Semicontinuous Functions

On $\mathbb{R}$ the condition for continuity of a real f at x_0 can be split into two parts: (i) $f(x_0) - \varepsilon < f(x)$ if $|x - x_0| < \delta$ and (ii) $f(x) < f(x_0) + \varepsilon$ if $|x - x_0| < \delta$. We thus define f as lower semicontinuous at x_0 if for every $\varepsilon > 0$ there exists $\delta > 0$ such that $f(x_0) - \varepsilon < f(x)$ if $|x - x_0| < \delta$. Similarly, for upper semicontinuous $f(x) < f(x_0) + \varepsilon$ if $|x - x_0| < \delta$. We define these for a topological space.

Definition 9. Let X be a topological space and $f : X \to \mathbb{R}$ a real function on X. Then f is called upper *semicontinuous* on X if $f^{-1}(-\infty, t)$ is open in X for every real t, i.e.

$$\{x \in X : f(x) < t\}$$

is open in X for every real t. A function f is called *lower semicontinuous* if $-f$ is upper semicontinuous.

Example 8. Let $X = \ell_\infty$ and define $f : \ell_\infty \to \mathbb{R} \cup \{\infty\}$ by

$$f(x) = \sum_{n=1}^{\infty} |x_n - x_{n+1}|$$

Then f is lower semicontinuous on ℓ_∞. In this example we allow f to take the value ∞ and by convention. If $f(\bar{x}) = \infty$ then f is called lower semicontinuous at $\bar{x}$ provided there is a neighbourhood of $\bar{x}$ in which $f(x) > \Delta$, for arbitrary $\Delta \in (0, \infty)$. To prove our assumption we consider just the case in which $f(\bar{x}) < \infty$ for $\bar{x} \in \ell_\infty$. The case $f(\bar{x}) = \infty$ is similar. Now there exists $N = N(\varepsilon)$ such that

$$\sum_{1}^{N} |\bar{x}_n - \bar{x}_{n+1}| > f(\bar{x}) - \frac{\varepsilon}{2}$$

Take $\delta = \varepsilon/4N$ and let $d(x, \bar{x}) < \delta$, where d is the metric in ℓ_∞. Then

$$\sum_{1}^{N} |\bar{x}_n - \bar{x}_{n+1}| \leq 2Nd(x, \bar{x}) + \sum_{1}^{N} |x_n - x_{n+1}|$$

so that

$$f(x) \geq \sum_{1}^{N} |x_n - x_{n+1}| > f(\bar{x}) - \varepsilon$$

whence f is lower semicontinuous at any $\bar{x} \in \ell_\infty$ for which $f(\bar{x}) < \infty$. It is not so difficult to show that f is not continuous on ℓ_∞. In fact we can construct a sequence of elements $x^{(i)} \in \ell_\infty$ such that $d(x^{(i)}, \theta) = 1/i$ for $i = 1, 2, \ldots$, where $\theta = (0, 0, \ldots)$ and such that $f(x^{(i)}) = 1$ for $i = 1, 2, \ldots$. Thus $x^{(i)} \to \theta$ in the metric of ℓ_∞ but $f(x^{(i)}) \nrightarrow f(\theta) = 0$, so that f is not continuous at θ.

4.9   Uniform Boundedness Principle

The next theorem is usually referred to as the uniform bounded principle which has important consequences and far-reaching applications.

Theorem 18. (Uniform boundeness principle). Let P be a collection of real lower semicontinuous functions p defined on the second category metric space X and suppose

$$p(x) \leq M(x) < \infty, \text{ each } x \in X, \text{ all } p \in P \tag{1}$$

Then there exists a sphere S in X and a constant M such that

$$p(x) \leq M, \text{ each } x \in S, \text{ all } p \in P \tag{2}$$

Proof. Let us consider the set

$$E(m, p) := \{x : p(x) \leq m\}$$

For each $p \in P$ and each positive integer m. This set is closed, since $E^c(m, p) := \{x : p(x) > m\}$ and p is lower semicontinuous. It follows that

$$E_m \equiv \cap \{E(m, p) : p \in P\}$$

being an intersection of closed sets, is closed. Now

$$X = \cup \{E_m : m = 1, 2, \ldots\} \tag{3}$$

For if $x \in X$ then $p(x) \leq M(x)$ for all $p \in P$ and so there is an integer $m(x)$ such that $p(x) \leq m(x)$ for all $p \in P$. This implies that $x \in E_{m(x)}$, which proves (3).

By hypothesis X is of the second category, so that (3) implies that at least one of the sets E_m, say E_M, is not nowhere dense (if all the

E_m were nowhere dense then the countable union of them would be of the first category). Since E_M is not nowhere dense we have that $\overline{E}_M$ contains some sphere S and the fact that E_M is closed implies that $S \subset E_M = \overline{E}_M$.

Finally, $x \in S$ implies $x \in E_M$, which implies $\varrho\ (x) \leq M$ for all $p \in P$. This proves (2).

Remark 4. The essential difference between (1) and (2) is that the inequality in (1) involves a bound $M(x)$ which depends on x but not on p and the inequality holds on the whole of X. In (2) the bond M is uniform depending neither on x nor on p, but this time the inequality holds only on a sphere in X, not necessarily on the whole of X. Usually the family P will be a sequence $\{p_n\}$ of functions.

Corollary 19. The theorem holds if we replace P by a collection F of continuous real functions f.

Corollary 20. The theorem holds if X second category is replaced by X complete.

Theorem 21. (Extension Theorem). Let (X, d_1) be a metric space and (Y, d_2) be a complete metric space. If f is uniformly continuous function from a subset A of X into Y, then f can be extended uniquely to a uniformly continuous function g from $\overline{A}$ into Y.

Proof. We prove the theorem in the following steps:

  (i) Existence of $g : \overline{A} \to Y$
  (ii) Uniform continuity of g
  (iii) Uniqueness of g

(i) Let $\{a_n\}$ be any convergent sequence in A converging to a point $x \in \overline{A}$. Also $\{a_n\}$ being convergent must be a Cauchy sequence and since f is uniformly continuous, its image $\{f(a_n)\}$ is a Cauchy sequence in Y. Again Y is given to be completed, the sequence $\{f(a_n)\}$ must be a convergent sequence in Y, and so there is a point $y \in Y$ such that $f(a_n) \to y\ (n \to \infty)$. Now we show that y depends only on x and not on the sequence $\{a_n\}$.

Let $\{b_n\}$ be another sequence in A such that $\{b_n\}$ converges to x, then by the triangle inequality in x, we have

$$d_1(a_n, b_n) \leq d_1\ (a_n, x) + d_1(x, b_n) \to 0\ (n \to \infty)$$

Now, by the uniform continuity of f

$$d_2\ (f(a_n), f(b_n)) \to 0\ (n \to \infty).$$

Again by the triangle inequality in Y, we have

$$d_2(f(b_n), y) \leq d_2(f(b_n), f(a_n)) + d_2(f(a_n), y) \to 0 \ (n \to \infty)$$

i.e. $\lim\limits_{n \to \infty} f(b_n) = y$. Hence, y does not depend on the sequence $\{a_n\}$ in A.

Thus if we define $y = g(x)$ then g extends f from A to $\overline{A}$ which can be as follows:

Let $x \in A$. Then $x \in \overline{A}$. Taking $a_n = x$ for every n, the sequence $\{a_n\}$ is a constant sequence in A and so $a_n \to x \ (n \to \infty)$. Thus $g(x) = \lim\limits_{n \to \infty} f(a_n)$. But since $f(a_n) = f(x)$, we have $\lim\limits_{n \to \infty} f(a_n) = f(x)$ for every $x \in A$. Therefore $f(x) = g(x)$ for every $x \in A$. Hence g extends f to $\overline{A}$.

(ii) Let $\varepsilon > 0$ be given. Then, since f is continuous, we find a $\delta > 0$ such that for all $a, b \in A$ we have

$$d_1(a, b) < \delta \implies d_2(f(a), f(b)) < \varepsilon \tag{1}$$

Let $x, y \in \overline{A}$ such that $d_1(x, y) < \delta$. Then there exist sequences $\{a_n\}$ and $\{b_n\}$ in A such that $a_n \to x$, and $b_n \to y \ (n \to \infty)$, respectively, That is, $d_1(a_n, x) \to 0$, and $d_1(b_n, y) \to 0$, $(n \to \infty)$.

Now, for $r = \dfrac{\delta - d_1(x, y)}{2} > 0$ there exists a positive integer n_0 (depending on r) such that

$$d_1(a_n, x) < r \text{ and } d_1(b_n, y) < r, \quad \forall \ n \geq n_0$$

Therefore

$$d_1(a_n, b_n) \leq d_1(a_n, x) + d_1(x, y) + d_1(y, b_n) < r + d_1(x, y) + r = \delta,$$

$\forall n \geq n_0$. From (1) it follows that

$$d_2(f(a_n), f(b_n)) < \varepsilon, \quad \forall \ n \geq n_0 \tag{2}$$

By definition of g we have

$$f(a_n) \to g(x) \quad \text{and} \quad f(b_n) \to g(y) \ (n \to \infty)$$

i.e. for each $\varepsilon > 0$, there exist positive integers m_1 and m_2 such that

$$d_2(f(a_n), g(x)) < \varepsilon/3, \quad \forall \ n \geq m_1$$

and $$d_2(f(b_n), g(x)) < \varepsilon/3, \quad \forall \ n \geq m_2$$

By the triangle inequality in Y

$$d_2(g(x), g(y)) \leq d_2(g(x), f(a_n)) + d_2(f(a_n), f(b_n)) + d_2(f(b_n), g(y))$$

Using (2) and by the definition of g, we get

$$d_2(g(x), g(y)) < \varepsilon/3 + \varepsilon/3 + \varepsilon/3 = \varepsilon, \ \forall \ n \geq m = \max \ \{n_0, m_1, m_2\}$$

Thus $d_1(x, y) < \delta \Rightarrow d_2(g(x), g(y)) < \varepsilon, \ \forall \ x, y \in \overline{A}$.

Hence g is uniformly continuous.

(iii) Let, if possible, there be another extension $h: \overline{A} \to Y$ of f to $\overline{A}$ such that h is uniformly continuous.

We have for all $x \in A$

$$g(x) = f(x) = h(x)$$

and for all $x \in \overline{A}$

$$g(x) = f(x) = h(x)$$

Hence, $g(x) = h(x), \ \forall \ x \in \overline{A}$, i.e. g is unique.

EXERCISES

1. Let d be a metric on a non-empty set X. Show that d is continuous on X.

2. Let (X, d_1) and (Y, d_2) be metric spaces, $f : X \to Y$ be a continuous function and $A \subset X$. Prove that the restriction of f to A, $g : A \to Y$ (defined by $g(a) = f(a)$, for $a \in A$) is continuous on A.

3. Prove that the composition of two uniformly continuous functions is also a uniformly continuous function.

4. Define f, g, h by $f(x) = 2x$, $g(x) = x^3$, $h(x) = 1/x$. Show that f is uniformly continuous on $\mathbb{R}$; g is continuous but not uniformly continuous on $\mathbb{R}$; g is uniformly continuous on $(0, 1)$; h is continuous but not uniformly continuous on $(0, 1)$.

5. Let X be a metric space and $f: \mathbb{N} \to X$. Prove that f is uniformly continuous on $\mathbb{N}$.

6. Give an example of a function $f : \mathbb{R} \to \mathbb{R}$ which is continuous but not open.

7. Let (X, d_1) and (Y, d_2) be metric spaces and $f : X \to Y$. Prove that f is continuous iff $f^{-1}(E^0) \subset (f^{-1}(E))^0$ for every $E \subset Y$.

8. Let (X, d_1) and (Y, d_2) be metric spaces and $f: X \to Y$ be one-to-one and onto. Prove that f is homeomorphism iff $f(E^0) = (f(E))^0$, for every $E \subset X$.

9. Let (X, d_1) and (Y, d_2) be metric spaces and $f: X \to Y$ be bijective. Prove that f is a homeomorphism iff $f(\overline{E}) = \overline{f(E)}$.

10. Let (X, d) be a metric space and $Y \subset X$. Prove that the inclusion map $i: Y \to X$ is continuous.

11. Let X and Y be metric spaces and A be a non-empty subset of X. If f and g are continuous functions from X into Y such that $f(x) = g(x)$, for every $x \in A$, show that $f(x) = g(x)$ for every $x \in \overline{A}$.

12. Let $x = \{x_n\} \in \ell_2$. Prove that f defined by $f(x) = \sum_{n=1}^{\infty} a_n x_n$, a_n's are real numbers, is a continuous real-valued function on ℓ_2.

13. Let f be a real-valued function on a metric space X. Prove that f is continuous on X iff the following sets $\{x : f(x) < c\}$ and $\{x : f(x) > c\}$ are open in X for every real number c.

14. Let (X, d) be a metric space such that $d(x, y) \leq 1$ for every $x, y \in X$, and let $\{x_n\}$ be a sequence in X. If $f(x) = \{d(x, x_n)\}$, $x \in X$, then prove that

(i) f is continuous function from X into H_∞, where H_∞ is the set of all real sequences $\{x_n\}$ such that $|x_n| \leq 1$, $\forall n \in \mathbb{N}$.

(ii) If the range set of $\{x_n\}$ is dense in X, then f is one-to-one.

15. Show that the mapping $f : \mathbb{N} \times \mathbb{N} \to \mathbb{Q}$ given by $(n_1, n_2) \to (n_1/n_2)$, is continuous when $\mathbb{Q}$ has the usual metric and $\mathbb{N} \times \mathbb{N}$ has the metric given by

$$\rho((n_1, n_2), (m_1, m_2)) = |n_1 m_2 - n_2 m_1| \cdot (n_2 m_2)^{-1}$$

16. Let (X, d) be a metric space with $x_0 \in X$. Define $f : X \to \mathbb{R}$ by $f(x) = d(x, x_0)$. Prove that f is uniformly continuous on X.

17. Let $f : X \to \mathbb{R}$, where X is a topological space. Prove that

(i) f is continuous iff it is both l.s.c. and u.s.c.

(ii) f, g l.s.c. (u.s.c.) imply $f + g$ l.s.c. (u.s.c.).

(iii) f_n u.s.c., $f_n \geq f_{n+1}$ on X, $f_n \to f$ on X imply f u.s.c.

18. Let T consist of $\varnothing$ and the complements of countable sets in $\mathbb{R}$. Prove that every function $f : (\mathbb{R}, T) \to (X, T')$, where (X, T') is any topological space, is sequentially continuous at each point of $\mathbb{R}$.

19. Let $C^\infty[0, T]$ be the set of all real-valued infinitely differentiable functions f defined on $0 \leq t \leq T$. Let D be the mapping of $C^\infty[0, T]$ into itself defined by $Df = \dfrac{df}{dt}$.

(i) Let d_1 be the sup-metric on $C^\infty[0, T]$. Is D a continuous mapping of $(C^\infty[0, T], d_1)$ into itself?

(ii) Define a metric d_2 on $C^\infty[0, T]$ by

$$d_2(x, y) = d_1(x, y) + d_1(Dx, Dy).$$

Is D a continuous mapping of $(C^\infty[0, T], d_2)$ into $(C^\infty[0, T], d_1)$?

20. Let X be a metric space and $f : \mathbb{N} \to X$ any function on the positive integers $\mathbb{N}$. Prove that f is uiformly continuous on $\mathbb{N}$.

21. Prove that if $f : X \to Y$ is an isometry from X onto Y, then f is uniformly continuous.

22. Let $Y = C[0, T]$ be given with the sup-metric $d(x, y)$. Let $X = (C[0, t], d) \times (C[0, t], d)$. Consider the mapping F of X into Y defined by $F(x) = x_1 x_2$, where $x = (x_1, x_2)$. Is F continuous? Is F uniformly continuous?

5

Compactness

Compactness is a topological property of metric spaces, that is, if the metric spaces X and Y are homeomorphic to one another, then X is compact iff Y is compact.

There are at least four equivalent ways of defining compactness in metric spaces. The first definition we choose is based on the aspect of compactness that is used most in applications, namely, sequential compactness. We shall see that the sequential compactness in a metric space is equivalent to three other forms of compactness.

First we define the concepts of ε-net and total boundedness.

5.1 Total Boundedness

Definition 1. Let A be a subset of a metric space (X, d). Given $\varepsilon > 0$, a subset B of A is said to be an ε-net of A if for each $x \in A$ there is $y \in B$ such that $d(x, y) < \varepsilon$. If the set B is finite, then the ε-net B of A is called *finite ε-net*.

Definition 2. A subset A of a metric space (X, d) is said to be *totally bounded* if given $\varepsilon > 0$, there exists a finite number of subsets A_1, A_2, ..., A_n of X such that $d(A_k) < \infty$ $(k = 1, 2, ..., n)$ and $A \subset \bigcup_{k=1}^{n} A_k$, that is, A can be covered by a finite number of subsets of X of finite diameter (see also Definition 5).

Thus, it follows that every subset of a totally bounded set is totally bounded.

Theorem 1. The subset A of a metric space (X, d) is totally bounded iff A has a finite ε-net for each $\varepsilon > 0$.

Proof. Suppose that A is totally bounded and $\varepsilon > 0$ be given. Then $A \subset \bigcup_{i=1}^{n} A_i$, where $d(A_i) < \varepsilon$ for each $i = 1, 2, \ldots, n$. We assume that each A_i is non-empty. If $a_i \in A_i$ for $i = 1, 2, , , , , n$, then we have to show that there is a set $B = \{a_1, a_2, \ldots, a_n\}$ which is ε-net of A. Let $x \in A$. Then $x \in A_i$ for some i. Since x and a_i are in A_i and $d(A_i) < \varepsilon$ for each i, we get $d(x, a_i) < \varepsilon$ for each $i = 1, 2, \ldots, n$. Hence A has a finite ε-net.

Conversely, let $B = \{x_1, x_2, \ldots, x_n\}$ be $\varepsilon/3$-net of A. Let $x \in A$, then there exists $x_i \in B$ such that $d(x, x_i) < \varepsilon/3$, that is, $x \in S_{\varepsilon/3}(x_i)$ and so $S_{\varepsilon/3}(x_1)$, $S_{\varepsilon/3}(x_2)$, ..., $S_{\varepsilon/3}(x_n)$ form a covering of A. If $y \in A$, then $x \in S_{\varepsilon/3}(x_k)$ for some $x_k \in B$. Now

$$d(x, y) < d(x, x_i) + d(x_i, x_k) + d(x_k, y) < \varepsilon/3 + \varepsilon/3 + \varepsilon/3 = \varepsilon$$

Hence, $S_{\varepsilon/3}(x_1)$, $S_{\varepsilon/3}(x_2)$, ..., $S_{\varepsilon/3}(x_n)$ is a covering of A by sets of diameter less than ε, i.e. A is totally bounded.

Theorem 2. Let (X, d) be a metric space. A subset A of X is totally bounded iff every sequence of points of A contains a Cauchy subsequence.

Proof. Let A be a totally bounded subset of X and $\{x_n\}$ be a sequence of points of A. Then, given $\varepsilon = 1$, A can be convered by a finite number of subsets of X of diameter less than 1, that is, $A \subset \overset{n}{\underset{i=1}{\cup}} A_i$ where $d(A_i) < 1$ for $i = 1, 2, \ldots, n$. Therefore, one of these sets, say, A_1, will contain infinitely many elements of $\{x_n\}$. For otherwise the sequence will be a finite sequence. Choose an element of A_1 say x_{n_1}. Now $A_1 \subset A$, and so A_1 is totally bounded. Hence, it can be covered by a finite number of subsets of A_1 of diameter less than $1/2$. One of the sets A_2 must contain x_n for infinitely many n. Let $n_2 > n_1$ such that $x_{n_2} \in A_2$. Since $A_2 \subset A_1$, $x_{n_2} \in A_1$. Proceeding in a similar manner, for any $k \in \mathbb{N}$ we can find a subset A_k of A_{k-1} with $d(A_k) < 1/k$ and $x_{n_k} \in A_k$. Since all $x_{n_1}, x_{n_2}, \ldots$ are in A_k and $d(A_k) < 1/k$, we see that $\{x_{n_k}\}_{k=1}^{\infty}$ is a Cauchy subsequence of $\{x_n\}$.

Conversely, suppose that every $\{x_n\}$ of points of some subset A of X contains a Cauchy sequence. We shall show that A is totally bounded. Suppose that A is not totally bounded. By Theorem 1, there exists some $\varepsilon > 0$ such that A contains no finite ε-net. Thus for $x_1 \in A$, the set $\{x_1\}$ is not ε-net of A. So there is a $x_2 \in A$ such that $d(x_1, x_2) > \varepsilon$. But then $\{x_1, x_2\}$ is not ε-net of A and so there exists, $x_3 \in A$ such that $d(x_1, x_3) > \varepsilon$ and $d(x_2, x_3) > \varepsilon$. Continuing in this way, we can construct a sequence $\{x_n\}$ of points of A such that $d(x_n, x_m) > \varepsilon$ for any m and n and $n \neq m$. Hence, $\{x_n\}$ has no Cauchy subsequence. This contradicts our hypothesis, and so A must be totally bounded.

Theorem 3. Let (X, d) be a metric space. Then every totally bounded set is bounded in (X, d) but converse need not be true.

Proof. Let B be an ε-net for a set A contained in (X, d). Then B contains a finite number of points $\{y_1, y_2, \ldots, y_n\}$. Now let $B = \max \{d(y_i, y_j) : 1 \le i, j \le n\}$, which is a finite number. We now claim that diam $(A) \le B + 2\varepsilon$. Indeed, if x_1 and x_2 are any two points in A, then there are two points in B, say, y_1 and y_2, such that

$$d(x_i, y_i) \le \varepsilon, \ i = 1, 2$$

Thus, we have

$$d(x_1, x_2) \le d(x_1, y_1) + d(y_1, y_2) + d(y_2, x_2) \le B + 2\varepsilon$$

Hence, diam $(A) \le B + 2\varepsilon$, i.e. A is bounded.
The converse part follows from Example 1 (Sec. 5.2).

Theorem 4. A totally bounded metric space is separable.

Proof. Let (X, d) be a totally bounded metric space. Then, by Theorem 1, X has a finite $1/n$-net for each $n = 1, 2, \ldots$, and we denote it by A_n. Clearly $\bigcup_{n=1}^{\infty} A_n$ is countable. Write $A = \bigcup_{n=1}^{\infty} A_n$. We have to show that $\overline{A} = X$. Let $A_n = \{a_1, a_2, \ldots, a_n\}$, since A_n is finite. Choose $n \in \mathbb{N}$ such that $1/n < r$ for $r > 0$. Then $X = \bigcup_{i=1}^{n} S_{1/n}(a_i)$.
Let $x \in X$. Then $x \in S_{1/n}(a_i)$ for some $i = 1, 2, \ldots, n$. Therefore

$$d(x, a_i) < 1/n < r$$

i.e. $a_i \in S_r(x)$.

Since $a_i \in A_n \subset A$, $S_r(x) \cap A \ne \varnothing$. Thus, $x \in \overline{A}$. Hence, $\overline{A} = X$.

5.2   Examples

1. Let us consider (l_2, d_2) with the metric

$$d_2(x, y) = \left\{ \sum_{k=1}^{\infty} |x_k - y_k|^2 \right\}^{1/2}$$

and let A be the set of all points in (l_2, d_2) such that $\sum_{k=1}^{\infty} |x_k|^2 \le 1$.
The set A, of course, is bounded. Since, by the Minkowski inequality

$$d_2(x, y) \le \left\{ \sum_{k=1}^{\infty} |x_k|^2 \right\}^{1/2} + \left\{ \sum_{k=1}^{\infty} |y_k|^2 \right\}^{1/2}$$

and we see that diam $(A) \le 2$. Yet the set A is not totally bounded. Consider the set $E = \{e_1, e_2, \ldots\}$ of points in A, where $e_1 = \{1, 0,$

$0, \ldots\}$, $e_2 = \{0, 1, 0, 0, \ldots\}$, and so on. We see that $d_2(e_k, e_j) = \sqrt{2}$ for $k \neq j$. If an ε-net exists for $\varepsilon = 1/2$, say B, there must be an appropriate finite set in A. But the closed balls $S_{1/2}[e_k]$ and $S_{1/2}[e_j]$ are disjoint for $k \neq j$. Thus, if the set B contains a point within distance $1/2$ of each e_k, then B must have at least one of its points of each closed ball $S_{1/2}[e_k]$. If follows that B must be at least countably infinite. Therefore, a finite ε-net for $\varepsilon = 1/2$ does not exist, and A is not totally bounded.

2. Consider the metric d on $\mathbb{R}^n$ as

$$d(x, y) = [|x_1 - y_1|^2 + \ldots + |x_n - y_n|^2]^{1/2}$$

and let A be the set in $(\mathbb{R}^n, d)$ of all points $x = (x_1, x_2, \ldots, x_n)$ such that $\sum_{i=1}^{n} |x_i|^2 \leq 1$.

The set A is, of course, bounded. Let us show that A contains a finite ε-net for each $\varepsilon > 0$. Let k be a positive integer such that $\sqrt{n} \leq \varepsilon k$. Let B be the set of all n-tuples $(y_1, y_2, \ldots, y_n)$ such that y_j ($j = 1, 2, \ldots, n$) can take only the values m/k (where m is an integer with $-k \leq m \leq k$) and $\sum_{j=1}^{n} |y_j|^2 \leq 1$. The set B is finite and $B \subset A$. It is apparent that for an arbitrary $x = (x_1, x_2, \ldots, x_n)$ in A there is a point $y = (y_1, y_2, \ldots, y_n)$ in B such that $|x_i - y_i| \leq 1/k$ or

$$d(x, y) = \left\{ \sum_{j=1}^{n} \left(\frac{1}{k} \right)^2 \right\}^{1/2} = \frac{\sqrt{n}}{k} \leq \varepsilon$$

Hence, B is a finite ε-net and ε being arbitrary, A is totally bounded.

3. Consider the metric space (l_2, d_2). Let A be the *Hilbert cube*, that is, the set of all points $x = \{x_n\}$ in (l_2, d_2) such that $|x_n| \leq 1/n$. It follows that

$$\lim_{N \to \infty} \sum_{n=N}^{\infty} |x_n|^2 = 0 \text{ uniformly on } A$$

that is, given $\varepsilon > 0$ there exists an integer $N = N(\varepsilon)$ such that

$$\sum_{n=N}^{\infty} |x_n|^2 \leq \sum_{n=N}^{\infty} \frac{1}{n^2} \leq \left(\frac{\varepsilon}{2} \right)^2$$

Carefully note that $N(\varepsilon)$ is independent of the point in A. Choose k to be an integer with $2\sqrt{N} \leq \varepsilon k$. Let B be the set of points $y = \{y_n\}$ in A such that $y_n = 0$ for $n \geq N$ and such that the values of y_i, $i = 1, 2, \ldots, N$, are restricted to the numbers m/k ($|m| \leq k$), in the spirit of Example 2. It is then simple to argue that for an arbitrary

$x = (x_1, x_2, \ldots)$ in A there is a point y in B such that $|x_i - y_i| \le 1/k$, $i = 1, 2, \ldots, N$, and

$$d_2^2(x, y) = \sum_{n=1}^{N} |x_n - y_n|^2 + \sum_{n=N+1}^{\infty} |x_n|^2 \le \frac{N}{k^2} + \frac{\varepsilon^2}{4} \le \varepsilon^2$$

If follows immediately that A is totally bounded.

4. Consider the space l_∞ with the sup-metric, and a subset $E = \{e_1, e_2, \ldots\}$ of l_∞. Then E is bounded but not totally bounded.

Since $d(e_n, e_m) = 1$, the set E is bounded. Further since $d(e_n, e_m) = 1$ for all m and n, the sequence $\{e_k\}_{k=1}^{\infty}$ has no Cauchy subsequence. Hence, by Theorem 2, E is not totally bounded.

5.3   Sequential Compactness

Definition 3. A metric space (X, d) is said to be *sequentially compact* if every sequence in (X, d) contains a convergent subsequence. A set $A \subset X$ is said to be sequentially compact if the subspace (A, d) is sequentially compact, that is, every sequence in A contains a subsequence that converges to a limit point in A.

We shall say that the sequence $\{x_1, x_2, \ldots\}$ *contains a convergent subsequence* if at least one of its subsequences is convergent, e.g. the sequence $\{1, 1/2, 3, 1/4, 5, 1/6, \ldots\}$ has $\{1, 1/2, 1/4, 1/6, \ldots\}$ as a convergent subsequence. Note that the sequence itself is not convergent, and it contains many subsequences, such as $\{1, 3, 5, 7, \ldots\}$ which are not convergent.

Remark 1. Roughly speaking, a sequentially compact metric space is so "crowded" that no matter how hard one tries to choose a sequence, an infinite number of the elements will always "pile up" arround at least one point in the metric space.

Theorem 5. A metric space (X, d) is sequentially compact iff it is totally bounded and complete.

Proof. Let X be totally bounded and complete. Then by Theorem 2, the sequence $\{x_n\}$ of points of X has a Cauchy subsequence $\{x_{n_k}\}$, since X is totally bounded. Further, since X is complete, every Cauchy sequence is convergent, i.e. $\{x_{n_k}\}$ converges to x. Hence, every sequence in X has a convergent subsequence, that is X is sequentially compact.

Conversely, let X be sequentially compact. Then every sequence in X has a convergent subsequence. Since every convergent sequence is a Cauchy sequence, every sequence has a Cauchy subsequence and so X is totally bounded by Theorem 2. To show the completeness, let $\{x_n\}$ be a Cauchy sequence in X. By hypothesis $\{x_n\}$ has a convergent subsequence $\{x_{n_k}\}$ converging to a point x in X. We know that if a Cauchy sequence has a convergent subsequence then the Cauchy sequence itself converges to the same point as its convergent subsequence. Hence $\{x_n\}$ converges to x in X. Therefore, X is complete.

Theorem 6. If A is a sequentially compact set in a metric space (X, d), then A is a closed set.

Proof. Let $x \in X$ be any limit point of A. Then there is a sequence $\{x_n\}$ in A converging to x. We have to show that $x \in A$. Since $\{x_n\}$ is a convergent sequence in A, it is a Cauchy sequence in A. Since A is sequentially compact, it is complete, by Theorem 5. Therefore, the sequence $\{x_n\}$ converges to a point in A and this point must be x. Hence $x \in A$. Thus A contains all its limit points. Hence, A is closed.

Example 5. (i) The set $(0, 1]$ is not sequentially compact in $\mathbb{R}$ since the sequence $\{1/n\}$ does not have a subsequence with a limit in $(0, 1]$.

(ii) The metric space $[0, 1]$ with the absolute value metric is sequentially compact.

(iii) Consider the set $[0, \infty)$ of $\mathbb{R}$. This is a closed subset of $\mathbb{R}$. But this is not sequentially compact, since it is not bounded in $\mathbb{R}$ and hence not totally bounded in $\mathbb{R}$.

Theorem 7. Let (X, d) be a sequentially compact metric space. A set $A \subset X$ is sequentially compact iff A is closed.

Proof. It follows from Theorem 6 that if A is sequentially compact, then it is closed. Now assume that A is closed. Let $\{x_n\}$ be a sequence in $A \subset X$. Since X is sequentially compact we can find a subsequence $\{x_{n_k}\}$ with limit x_0 in X. Since A is closed by the Closed Set Theorem (see Exercise 28, Chapter 2), $x_0 \in A$. Hence, every sequence in A has a convergent subsequence, i.e. A is sequentially compact.

Corollary 8. Any sequentially compact subset A of a metric space (X, d) is closed and bounded.

Proof. By Theorem 6, A is closed. Since A is sequentially compact, it is totally bounded by Theorem 5. Since a totally bounded set is bounded, A is bounded.

Example 6. Every finite subset E of any metric space (X, d) is sequentially compact.

We known that every finite subset E of a metric space is totally bounded. If $\{x_k\}_1^\infty$, is any sequence in E, then at least one of the elements x_0, say, must appear infinite number of times in the sequence. Hence $\{x_0, x_1, \ldots\}$ is a subsequence of $\{x_n\}$. This subsequence is convergent to x_0 in E. Hence, E is totally bounded and complete. By Theorem 5, E is sequentially compact.

5.4　Bolzano-Weierstrass Property

The well known Bolzano-Weierstrass theorem, "every closed and bounded infinite subset in $\mathbb{R}$ has a limit point", is no longer true in general metric spaces. For example, if we consider the metric space (X, d), where $X = (0, 1]$ with the usual metric d, and the infinite subset $A = \{1, 1/2, 1/3, \ldots\}$ of X. Here 0 is the only limit point of A which is not in X, This motivates the following definition.

Definition 4. A metric space (X, d) is said to possess the *Bolzano-Weierstrass Property* (BWP) if every infinite subset of (X, d) has at least one limit point. A set A in (X, d) is said to possess the BWP if the space (A, d) has this property.

Theorem 9. A metric space is sequentially compact iff it has the BWP.

Proof. Let (X, d) be a sequentially compact metric space. Let A be an infinite subset of X. We can obtain a sequence $\{x_n\}$ of distinct points of A. Since X is sequentially compact, $\{x_n\}$ has a subsequence $\{x_{n_k}\}$, say which converges to a point x in X.

Let $\varepsilon > 0$ be given. Since $x_{n_k} \to x (k \to \infty)$, there exists an integer $N > 0$ such that $d(x_{n_k}, x) < \varepsilon$ for all $k \geq N$, i.e. $x_{n_k} \in S_\varepsilon(x) \ \forall \ k \geq N$. But $x_{n_k} \neq x$ for all $k \geq N$, because $\{x_n\}$ consists of district elements form A. Thus $S_\varepsilon(x)$ contains an element x_{n_k}, for some $k \geq N$, of A, different from x. Hence x is a limit point of A. Therefore, X has the BWP.

Conversely, let X have the BWP. Let $\{x_n\}$ be an arbitrary sequence in X and A be the range of the sequence $\{x_n\}$. There are two cases:

(i) A is finite. This is possible only in one situation when at least one of the elements of A occurs infinitely many times in the sequence $\{x_n\}$, and so we get a constant subsequence converges to that element.

(ii) A is infinite. In this case by the BWP, it has at least one limit point, say, x in X. Then $S_1(x)$ must have an infinite number of elements of A. Let us choose one of such elements x_{n_1}, say, different from x_0. Again $S_{1/2}(x)$ must have infinite number of elements of A and choose one of such elements x_{n_2} $(n_2 > n_1)$. Continuing this process, we can get a subsequence $\{x_{n_k}\}$ of

$\{x_n\}$ such that $x_{n_k} \in S_{1/k}(x)$ for every k, that is

$$d(x_{n_k}, x) < \frac{1}{k} \to 0 \quad \text{as} \quad k \to \infty$$

Thus $x_{n_k} \to x$. Hence X is sequentially compact.

5.5 Heine-Borel Compactness

There is yet another way of characterizing sequentially compact metric spaces which is often useful, i.e. Heine-Borel compactness and we use the term "compactness" for Heine-Borel compactness.

First we need the following definition.

Definition 5. Let (X, d) be a metric space. A family $\mathscr{F} = \{G_\alpha : \alpha \in \Lambda\}$ of subsets of X is said to be a *cover* of X if $X \subset \bigcup_{\alpha \in \Lambda} G_\alpha$. If each G_α is open, then $\mathscr{F}$ is called an *open covering* of X. Further, if a finite subfamily of $\mathscr{F}$ covers X, then the finite subfamily is called a *finite subcovering* of X.

Example 7. Let $\mathbb{R}$ be the usual metric space.

(i) Let $\mathscr{F} = \{(-n, n) : n \in \mathbb{N}\}$ and

$$\mathscr{F}' = \{(-2n, 2n) : n \in \mathbb{N}\}.$$

Then $\mathscr{F}$ is an open cover of $\mathbb{R}$ and $\mathscr{F}'$ is a subcover of $\mathscr{F}$.

(ii) Let $\mathscr{F} = \{(a - 1/n, b + 1/n) : n \in \mathbb{N}\}$.

Then $$[a, b] \subset \bigcup_{n=1}^{\infty} \left(a - \frac{1}{n}, b + \frac{1}{n}\right)$$

Hence, $\mathscr{F}$ is an open covering of $[a, b]$. Since any one open interval of the covering contains $[a, b]$, $\mathscr{F}$ contains a finite subcovering.

Definition 6. A metric space (X, d) is said to be *compact* if every open covering of X contains a finite open subcovering. A set A in a metric space (X, d) is said to be *compact* if the metric space (A, d) is compact.

Remark 2. (i) The interesting point about compact spaces is that even open coverings made up of 'very small' open sets contain finite subcoverings.

(ii) This version of compactness is probably the hardest to understand. However, it is this version that is really the most fundamental. The reason for this is that compactness can easily be generalized to topological spaces that are not metrizable.

Example 8. (i) The usual metric space $\mathbb{R}$ is not compact. Of course

$\mathcal{F} = \{(-n, n) : n \in \mathbb{N}\}$ is an open cover of $\mathbb{R}$ but it has no finite subcollection which also covers $\mathbb{R}$.

(ii) A finite set in any metric space (X, d) is compact.

(iii) The discrete metric space X_d, where X is an infinite set, is not compact.

Example 9. The subset $E = \{x = \{x_k\} : |x_k| < 1/k\}$ of l_∞ is compact.

Let $\{x^{(n)}\}$ be any infinite sequence in l_∞ where $x^{(n)} = \{x_1^{(n)}, x_2^{(n)}, \ldots\}$. By hypothesis the set of all first coordinates belongs to $[-1, 1]$. The sequence formed by the set of all first coordinates is bounded. Hence, $\{x_1^{(n)}\}_{n=1}^\infty$ contains a convergent subsequence converging to y_1, say. Hence, there is a subsequence of $\{x^{(n)}\}$ whose first coordinates will converge to y_1. Consider this subsequence of $\{x^{(n)}\}$ having the convergent first coordinates. In this subsequence the second coordinates are bounded and they lie in $[-1/2, 1/2]$. Hence, these second coordinates will have a convergent subsequence having the limit y_2. Now consider the subsequence of $\{x^{(n)}\}$ in which both the first and second coordinates sequences converge to y_1 and y_2. Proceeding in similar way, we obtain a subsequence of $\{x^{(n)}\}$ which converges to the limit $\{y_1, y_2, \ldots\}$ such that $|y_k| < 1$. Thus, every infinite sequence in E has a convergent subsequence. Hence, the subset E is compact.

Example 10. Let $E = [0, 1]$ with the discrete metric. The set $A = \{1, 1/2, 1/3, \ldots\}$ is bounded and closed but not compact.

A is obviously bounded. Since it has no limit points, it is closed. Now consider the open covering $G = \{S_1(1/n) : n \in \mathbb{N}\}$. This open covering does not contain any finite subcovering. For if it contains a finite subcovering, it contains only a finite number of points of A. Hence A is not compact.

Theorem 10. A metric space (X, d) is compact iff it is sequentially compact.

Proof. Let X be compact. We shall show that every sequence of points $\{x_n\}$ in X has a convergent subsequence converging to a point x in X. First, let us suppose that about each point x in X, there is an open sphere S_x which contains x_n for finitely many values of n. The family of all such open spheres S_x is an open covering of X. Since X is compact, it should be covered by a finite number of these S_x. This is not possible, since each S_x contains only a finite number of x_n's, the union of any finite number of open spheres S_x cannot contain all the x_n's. Hence, there must be some point x in X such that every open sphere about x contains x_n for infinitely many values of n. Thus, there exists $n_1 \in \mathbb{N}$ such that $x_{n_1} \in S_1(x)$ and similarly there exists $n_2 > n_1$ such that $x_{n_2} \in S_{1/2}(x)$. In general, there exists $n_k > n_{k-1}$ such that $x_{n_k} \in S_{1/k}(x)$. This subsequence $\{x_{n_k}\}$ of $\{x_n\}$ converges to x in X. Hence, X is sequentially compact.

Conversely, let X be sequentially compact. Then by Theorem 5, it is totally bounded and complete. On contrary, suppose that X is not compact. Then there exists at least one covering $\mathscr{T}$ of X such that $\mathscr{T}$ does not have any finite subcovering of X. Since X is totally bounded, it can be written as the union of a finite number of bounded subsets of X each of whose diameter is less than 1. Then one of these subsets A_1, say, cannot be covered by a finite number of sets in $\mathscr{T}$. If this is not so, X can be covered by a finite number of sets from $\mathscr{T}$ contradicting our assumption.

Now, since $d(\overline{A_1}) = d(A_1)$, $\overline{A_1}$ is a closed subset of X whose diameter is less than 1 and which cannot be covered by a finite number of sets in $\mathscr{T}$. Now $\overline{A_1}$ is sequentially compact and hence it is totally bounded by Theorem 5. Hence, as above, there is a subset A_2 of $\overline{A_1}$ with $d(A_2) < 1/2$ and $\overline{A_2}$ cannot be covered by a finite number of sets from $\mathscr{T}$. Thus $\overline{A_2} \subset \overline{A_1}$ with $d(\overline{A_2}) < 1/2$ and $\overline{A_2}$ cannot be covered by a finite number of sets in $\mathscr{T}$. Proceeding in this manner, we find that there exists a sequence of sets $\{\overline{A_n}\}$ of X such that

(i) $\overline{A_1} \supset \overline{A_2} \supset \overline{A_3} \supset \ldots \supset \overline{A_n} \ldots$

(ii) $d(\overline{A_n}) < 1/n$, $n = 1, 2, \ldots$

(iii) No finite number of sets of $\mathscr{T}$ can cover $\overline{A_n}$ for any $n \in \mathbb{N}$.

Since X is complete, by Cantor's Intersection theorem, $x \in \bigcap_{n=1}^{\infty} \overline{A_n}$. Since $\mathscr{T}$ is a covering of X, there is an open set G in $\mathscr{T}$ such that $x \in G$. This implies that there is an open sphere $S_r(x) \subset G$ for some $r > 0$. If the positive m is such that $1/m < r$, then $d(A_m) < 1/m \le r$. Since $x \in \overline{A_m}$, we have $\overline{A_m} \subset S_r(x) \subset G$. Hence, G alone covers $\overline{A_m}$. Thus, a finite subcovering of only one set G covers $\overline{A_m}$. But this is a contradiction to our assumption that no finite subcovering of $\mathscr{T}$ is a covering for $\overline{A_m}$. Hence, X is compact.

5.6 Finite Intersection Property

Section 5.5 showed that the compactness depends upon the family of open sets. Now we give an equivalent characterization of compact space by using closed sets in the metric space.

Definition 7. A family $\mathscr{F}$ of subsets of a set X is said to have the *finite intersection property* (f.i.p.), if the intersection of any finite number of sets in $\mathscr{F}$ is never empty, that is, for every finite collection $F_1, F_2, \ldots, F_n \in \mathscr{F}$,

$$\bigcap_{i=1}^{n} F_i \neq \varnothing$$

Example 11. (i) The family of all closed intervals

$$\mathscr{F} = \{[-1/n, 1/n] : n \in \mathbb{N}\}$$

has the finite intersection property.

(ii) The family of open intervals

$$\mathscr{F} = \{(0, 1/n) : n \in \mathbb{N}\}$$

has finite intersection property, since

$$\bigcap_{i=1}^{n} (0, b_i) = (0, b) \neq \varnothing$$

where $b = \min \{b_1, \ldots, b_n\} > 0$. Note that $\mathscr{F}$ itself has empty intersection.

Theorem 11. A metric space X is compact iff every family $\mathscr{F}$ of closed subsets of X with finite intersection property, has itself non-empty intersection, that is

$$\bigcap_{F \in \mathscr{F}} F \neq \varnothing \tag{1}$$

Proof. Let X be a compact metric space and $\mathscr{F}$ be a family of closed subsets of X with finite intersection property. We have to show that (1) holds.

Suppose, if possible.

$$\bigcap_{F \in \mathscr{F}} F = \varnothing$$

Then we can write $X = X - \bigcap_{F \in \mathscr{F}} F$. For our convenience, we shall write $\cap$ for $\bigcap_{F \in \mathscr{F}}$. By DeMorgan law

$$X - \cap F = \cup (X - F)$$

Therefore, $X = \cup (X - F)$. Since each F is closed, $X - F$ is open and so $\{(X - F)\}_{F \in \mathscr{F}}$ forms an open covering for X. Since X is compact, there exists a finite number of sets $F_1, F_2,\ldots, F_n$ such that $X = \bigcup_{i=1}^{n} (X - F_i)$ which implies that $X = X - \bigcap_{i=1}^{n} F_i$. Hence, $\bigcap_{i=1}^{n} F_i = \varnothing$.

This contradicts that F has the finite intersection property. Hence (1) holds.

Conversely, let every collection of closed subsets of X with finite intersection property have non-empty intersection. We shall show that X is compact. Let $\{G_i\}$ be an open covering of X so that $X = \cup\, G_i$ which gives $X - \cup\, G_i = \varnothing$. This implies, by DeMorgan law, that $\cap\, (X - G_i) = \varnothing$. Thus $\{(X - G_i)\}$ is a collection of closed sets with empty intersection and so by hypothesis this collection does not have finite intersection property. Hence, there exists a finite number of sets

$$X - G_1, X - G_2, \ldots, X - G_n$$

such that

$$\bigcap_{i=1}^{n} (X - G_i) = \varnothing$$

which implies that

$$X - \bigcup_{i=1}^{n} G_i = \varnothing$$

So that we have

$$X = \bigcup_{i=1}^{n} G_i$$

that is, the open covering $\{G_i\}$ has a finite subcovering and hence X is compact.

Now, we summarize the characterizations about a compact metric space as follows.

Theorem 12. Let (X, d) be a metric space. Then the following statements are equivalent:

  (a) X is compact
  (b) X is sequentially compact
  (c) X is totally bounded and complete
  (d) X has the Bolzano-Weierstrass Property
  (e) Every collection of closed subsets of X having f.i.p. has nonempty intersection.

Proof. It follows from Theorems 5, 9, 10 and 11.

5.7 Relation between Compactness and Continuity

This section exibits some properties of compact metric spaces for a continuous mapping.

Theorem 13. Let X and Y be two metric spaces and $f : X \to Y$ be a continuous mapping. If $A \subset X$ is compact in X, then $f(A)$ is compact in Y, that is, the continuous image of a compact metric space is compact.

Proof. Let A be compact. Then A is sequentially compact. Consider an arbitrary sequence $\{y_n\}$ in $f(A)$. For each y_n, we can choose $x_n \in A$ such that $f(x_n) = y_n$. Thus, we get a sequence $\{x_n\}$ in A. Since A is sequentially compact, $\{x_n\}$ has a convergent subsequence $\{x_{n_k}\}$ and so the continuity of f implies that the corresponding subsequence $\{f(x_{r_k})\} = \{y_{n_k}\}$ of $\{y_n\}$ is convergent. Hence $f(A)$ is sequentially compact and hence compact.

Corollary 14. Let f be continuous function from the compact metric space X into a metric space Y. Then the image $f(X)$ is a bounded and closed subset of Y.

Proof. By above theorem, $f(X)$ is compact. Since a compact subset of a metric space is closed, $f(X)$ is closed. Further, since $f(X)$ is compact, it is totally bounded. Hence $f(X)$ is bounded.

Example 12. Let $X = \{1, 1/2, 1/3, \ldots\}$ with the usual metric restricted on subsets of $\mathbb{R}$. Let $f : X \to \mathbb{N}$ defined by $f(1/n) = n$. Since the usual metric restricted to X is discrete, f is continuous. Further $\{1/n\}$ is a Cauchy sequence in X but $\{n\}$ is not a Cauchy sequence in $\mathbb{N}$. Thus, continuous image of a Cauchy sequence is not in general a Cauchy sequence.

Remark 3. By above example, we see that there are some properties of metric spaces which are not preserved by continuous functions.

Theorem 15. If f is one-to-one and continuous function from the compact metric space X onto a metric space Y, then f^{-1} is continuous on Y and hence f is a homeomorphism of X onto Y.

Proof. It is enough to show that f^{-1} is continuous on Y. Since $f : X \to Y$ is bijective, its inverse image $g = f^{-1}$ is well-defined from Y onto X. If F is a closed set in X, then F is compact in X by Theorem 7. Moreover $f(F)$ is compact in Y, since f is continuous. Therefore, again by Theorem 7, $f(F)$ is closed in Y. Hence f^{-1} is continuous, since for any closed set F in X, $g^{-1}(F) = f(F)$ is closed in Y. Therefore, f is a homeomorphism of X onto Y.

Note. In the above theorem, compactness on X cannot be dropped.

Example 13. Let $X = [0, 1)$ with the usual metric of $\mathbb{R}$ and let f be the mapping given by $f(x) = e^{2\pi i x}$, $0 < x < 1$. f is a one-to-one mapping of $[0, 1)$ into the circumference of the unit circle T in $\mathbb{R}^2$, f^{-1} is not continuous at the point $f(0)$.

First note that X is not compact. Consider the sequence $x_n = 1 - 1/n$. Then $f(x_n) = e^{-2\pi i/n}$. Hence, $f(x_n) \to 1$ as $n \to \infty$. But $f(0) = 1$ so that $f(x_n) \to f(0)$. But the corresponding preimage sequence $\{1 - 1/n\}$ does not converge in X. Thus, the compactness condition on X in the above theorem cannot be dropped.

5.8 Relation between Compactness and Uniform Continuity

We will see that on a compact metric space a function is continuous iff it is uniformly continuous. Since uniform continuity implies continuity, we have to prove the converse part only.

Theorem 16. Let (X, d_1) be a compact metric space. If f is a continuous function from X into a metric space (Y, d_2), then f is uniformly continuous on X.

Proof. Since f is continuous at each point of X, for a given $\varepsilon > 0$, there exists $\delta = \delta(a) > 0$ such that, for each $a \in X$

$$d_2(f(x), f(a)) < \varepsilon/2 \text{ whenever } d_1(x, a) < \delta \qquad (1)$$

Since X is compact, the family of all open spheres $S_{r/2}(a)$ for all $a \in X$ is an open covering of X and has a finite subcovering. Hence

$$S_{r_1/2}(a_1), S_{r_2/2}(a_2), \ldots, S_{r_n/2}(a_n)$$

form a covering of X.

Let $\delta = \min\left\{\frac{1}{2}r_1, \frac{1}{2}r_2, \ldots, \frac{1}{2}r_n\right\}$. Now for any point $a \in X$, we have $a \in S_{r_k/2}(a_k)$ for some $k = 1, 2, \ldots, n$ and so $d_1(a, a_k) < \frac{1}{2}r_k$. Now if $d_1(x, a) < \delta$, then $d_1(x, a) < \frac{1}{2}r_k$, since $\delta < \frac{1}{2}r_k$, $k = 1, 2, \ldots, n$. Hence, we get

$$d_1(x, a_k) \leq d_1(x, a) + d_1(a, a_k) < r_k \qquad (2)$$

Using (1), we obtain

$$d_2(f(x), f(a_k)) < \frac{\varepsilon}{2} \quad \text{and} \quad d_2(f(a), f(a_k)) < \frac{\varepsilon}{2} \qquad (3)$$

Now, by (2) and (3), we have

$$d_2(f(x), f(a)) \leq d_2(f(x), f(a_k)) + d_2(f(a), f(a_k)) < \varepsilon$$

Thus for our $\delta = \min \left\{ \frac{1}{2} r_1, \ldots, \frac{1}{2} r_k \right\}$, for all $a \in X$,

$$d_2(f(x), f(a)) < \varepsilon \text{ wherever } d_1(x, a) < \delta$$

i.e. this δ works in the definition of uniform continuity. Hence, f is uniformly continuous on X.

Corollary 17. If the real valued function f is continuous on the closed and bounded interval $[a, b]$, then f is uniformly continuous on $[a, b]$.

Remark 4. The following example shows that the condition of compactness cannot be dropped.

Example 14. The function $f : (0, 1) \to \mathbb{R}$ defined by $f(x) = 1/x$ is not uniformly continuous.

It is easy to check that f is continuous on $(0, 1)$.

Let $\varepsilon = \frac{1}{2}$. Choose $n > 1$ such that $1/n < \delta$. Let $x_1 = 1/n$ and $x_2 = 1/(n + 1)$. Then $x_1, x_2 \in (0, 1)$, and

$$|x_1 - x_2| = |1/n - 1/(n + 1)| < 1/n < \delta$$

But $|f(x_1) - f(x_2)| = |1/x_1 - 1/x_2| = |n - n - 1| = 1 > \varepsilon$

Hence f is not uniformly continuous on $(0, 1)$ which is not compact.

5.9   Relation between Continuity and Uniform Convergence

The following Dini's theorem gives some special conditions under which a sequence of functions on a metric space converges uniformly.

Theorem 18. Let $\{f_n\}$ be a monotonic increasing sequence of continuous real valued functions on a compact metric space (X, d). If $\{f_n\}$ converges pointwise on X to a continuous function f, then $\{f_n\}$ converges uniformly on X.

Proof. Let us write $g_n = f - f_n$ for each n. Since $\{f_n\}$ is a monotonic increasing sequence, $g_n > 0$ and $\{g_n\}$ is a monotonic decreasing sequence of continuous functions bounded below by 0. Further, we get $g_n(x) \to 0$ $(n \to \infty)$ for every $x \in X$, since $\{f_n\}$ converges pointwise to f on X. We have to show that $\{g_n\}$ converges to zero uniformly on X.

Let $\varepsilon > 0$ be given and let $x \in X$. Since $g_n(x) \to 0$ as $n \to \infty$, there exists an $m(x) \in \mathbb{N}$ such that $g_{m(x)}(x) < \varepsilon$. Since $g_{m(x)}$ is continuous at x, there exists an open sphere $S(x)$ such that $g_{m(x)}(y) < \varepsilon$ for every $y \in S(x)$. Now $\{S(x) : x \in X\}$ is an open covering of X. Since X is compact, this open covering has a finite subcovering. Let it be $S(x_1), S(x_2), \ldots, S(x_k)$. Having found out $x_1, x_2, \ldots, x_k$, we can find the corresponding $m(x_1), m(x_2), \ldots, m(x_k)$. Let $m = \max \{m(x_1), m(x_2), \ldots, m(x_k)\}$ and y be any point in X. Then $y \in S(x_i)$ for some $i = 1, 2, \ldots, k$. Hence, we get $g_{m(x_i)}(y) < \varepsilon$ by using the continuity of $g_{m(x_i)}$. Since $\{g_n\}$ is a decreasing sequence bounded below by 0 and $m(x_i) < m$, we get $g_m(y) < g_{m(x_i)}(y) < \varepsilon$ and $0 < g_m(y) < \varepsilon$ for all $y \in X$. Thus, we get $0 \leq g_n(y) < \varepsilon$ for all $n > m$, and all $y \in X$. Hence, $\{g_n\}$ converges uniformly to 0 on X.

5.10 Arzela-Ascoli's Theorem

Let (X, d_1) be a compact metric space and let (Y, d_2) be a complete metric space. Form the space $C = C(X, Y)$ of continuous functions defined on X with range in Y. If $f, g \in C$, define a metric ρ by

$$\rho(f, g) = \sup \{d_2(f(x), g(x)) : x \in X\}$$

Let A be a collection of functions from C. We seek conditions on A that will ensure that the closure $\overline{A}$ is a compact set in (C, ρ). First we define some related terms.

Definition 8. Let (X, d) be a metric space. A set $A \subset X$ is said to be *relatively compact* or *conditionally compact* if the closure of A is compact in (X, d).

Theorem 19. Let (X, d) be a metric space and $A \subset X$. Then A is relatively compact iff every sequence in A has a subsequence that converges to a point in X.

Proof. Let $\overline{A}$ be compact. Then every sequence in $A \subset \overline{A}$ has a subsequence that converges to a point in $\overline{A} \subset X$.

Conversely, let $\{x_n\}$ be a sequence in $\overline{A}$. It follows from the definition of the closure $\overline{A}$ that there is a sequence $\{y_n\}$ in A such that $d(x_n, y_n) \leq 1/n$. Choose a subsequence $\{y_{n'}\}$ of $\{y_n\}$ such that $\{y_{n'}\}$ converges. Let $z = \lim y_{n'}$. It is clear that $z \in \overline{A}$. Since

$$d(z, x_{n'}) \leq d(z, y_{n'}) + d(y_{n'}, x_{n'}) \to 0$$

we see that $z = \lim x_{n'}$. Hence $\overline{A}$ is sequentially compact, hence compact.

Definition 9. A family of functions A in C is said to be *pointwise compact* if for each $x \in X$ the set $\{f(x) : f \in A\}$ is relatively compact in (Y, d_2).

Definition 10. A family of functions A in C is said to be *equi-*

continuous if for each $x \in X$ and $\varepsilon > 0$ there is a $\delta > 0$ such that $d_2(f(x), f(x')) < \varepsilon$ for every f in A, where $d_1(x, x') < \delta$.

Note that δ depends on ε and x but not on f. If δ can be chosen independent of x as well, the family is said to be *uniformly equicontinuous*.

Theorem 20. (Arzela-Ascoli). Let A be a set in (C, ρ). Then A is relatively compact iff the family A is pointwise compact and equicontinuous.

Proof. Let A be relatively compact. In order to show that A is pointwise compact we fix x and choose a sequence $\{y_n\}$ in $\{f(x) : f \in A\}$, i.e. $y_n = f_n(x)$ for some $f_n \in A \subset \overline{A}$. Since $\overline{A}$ is compact, i.e. we can find a subsequence $\{f_{n'}\}$ that converges in (C, ρ), say $y_{n'} \to f$ as $n' \to \infty$. If we let $y = f(x)$, then we see that $y_{n'} \to y$ as $n' \to \infty$. Hence A is pointwise compact. (By the same argument we can show that $\overline{A}$ is pointwise compact).

Next we show that A is equi-continuous. Since $\overline{A}$ is compact, it is totally bounded. Let $\{f_1, f_2, \ldots, f_N\}$ be an ε-net for $\overline{A}$. If f is any point in $\overline{A}$ then there is an f_i in the ε-net such that

$$d_2(f(x), f_i(x)) \leq \rho(f, f_i) < \varepsilon$$

It follows that

$$d_2(f(x_0), f(x')) < 2\varepsilon + d_2(f_i(x_0), f_i(x')) \qquad (1)$$

Since each of the functions $\{f_1, \ldots, f_N\}$ is continuous at x_0, i.e. we can find a $\delta = \delta(x_0, \varepsilon) > 0$ such that

$$d_2(f_i(x_0), f_i(x')) < \varepsilon, \ 1 \leq i \leq N \qquad (2)$$

whenever $d_1(x_0, x') < \delta$. By combining (1) and (2), we see that $\overline{A}$ is equi-continuous. Hence A is equi-continuous.

Conversely, suppose A is pointwise bounded and equi-continuous. Since X is compact, we know that it is separable. Let $D = \{x_n\}$ be a countable dense set in X. We use Theorem 19 to show that $\overline{A}$ is compact. Let $\{f_n\}$ be a sequence in A. Since $\{f_n(x_1)\}$ lies in a compact set in Y, we can find a convergent subsequence, which we denote by $\{f_n^{(1)}\}$. Since $\{f_n^{(1)}(x_2)\}$ lies in a compact set in Y, we can find a convergent subsequence, which we denote by $\{f_n^{(2)}\}$. Continuing with $x_3, x_4, \ldots$, we construct a family

$$\{f_n^{(1)}\}, \{f_n^{(2)}\}, \{f_n^{(3)}\}, \ldots$$

of sequences, each a subsequence of the preceding, and with the

5. Show that every finite subset of a metric space is totally bounded.

6. Prove that a set E in the usual metric space $\mathbb{R}$ is compact iff it is closed and bounded.

7. If A and B are subsets of $\mathbb{R}$, prove that $A \times B$ is a compact subset of $\mathbb{R}^2$.

8. If f is a continuous real valued function on $[a, b]$, prove that the graph of f is a compact subset of $\mathbb{R}^2$.

9. Find an open covering for $[a, b]$ which contains a finite subcovering in the usual metric of $\mathbb{R}$.

10. Find an open covering of $(0, 1)$ which does not contain a finite subcovering.

11. Prove that compact metric space is compact iff every homeomorphic image of it is complete.

12. Prove that a compact metric space is separable.

13. Give an example of a complete metric space which is not compact.

14. Find an example of a closed and bounded subset of (l_2, d_2) which is not compact.

15. Let (X, d) be a sequentially compact metric space and let $\{M_n\}$ be a decreasing sequence of nonempty closed sets. Prove that $\bigcap_{n=1}^{\infty} M_n$ is nonempty.

16. Let (X, d_1) and (Y, d_2) be homeomorphic metric spaces. Show that (X, d_1) is compact iff (Y, d_2) is compact.

17. Let (X_1, d_1) and (X_2, d_2) be compact metric spaces. Prove that the product space (X, d), where $X = X_1 \times X_2$ and $d(x, y) = d_1(x_1, y_1) + d_2(x_2, y_2)$, is compact.

18. Let f be continuous real-valued function defined on a compact metric space (X, d). Prove that f is bounded, i.e.

$$M = \sup \{f(x) : x \in X\} \text{ and } m = \inf \{f(x) : x \in X\}$$

are finite.

19. Discuss the concept of sequential compactness in a semi-metric space.

20. Prove that any closed sphere in $\mathbb{R}^2$ is compact.

21. Find the compact subset of X_d.

22. Prove that a metric space isometric with a compact metric space is necessarily compact.

23. Let $A = \mathbb{N} \times \mathbb{N}$, and

$$F_{m,n} = \{(x, y) : x, y \in \mathbb{R}, \text{ and } |x| > m, |y| > n\}$$

Show that $\{F_{m,n}\}$ has the finite intersection property, and further show that $\cap \{F_{m,n}\} = \varnothing$.

24. Let (X, d) be a metric space and let A and B be subsets of X. Prove that if A is closed and B is compact and $d(A, B) = 0$, then $A \cap B \neq \varnothing$.

25. Show that a subset of $\mathbb{R}^n$ is bounded iff it is totally bounded.

26. Prove that boundedness and total boundedness are equivalent in Euclidean spaces.

6

Connectedness

Intuitively a connected set is a single piece, i.e. it does not consist of two or more separated pieces like the interval [0, 1] in $\mathbb{R}$ or a rectangle in $\mathbb{R}^2$. The sets of the type [1, 2] $\cup$ [3, 4] or $\mathbb{R} - \{0\}$ are not connected sets. Thus, we can say that a set is connected if it has no separation or if a set has separation it is not connected, i.e. disconnected. We shall define connectedness through the concept of separation.

6.1   Separated Sets

Definition 1. Let A be a subset of a metric space (X, d). A pair of subsets A_1 and A_2 of A is said to be a *separation* of A if: (i) $A_1 \neq \varnothing$, $A_2 \neq \varnothing$, $A = A_1 \cup A_2$, (ii) $A_1 \cap A_2 = \varnothing$, and (iii) $\overline{A_1}$ and $\overline{A_2}$ are closures of A_1 and A_2 in (X, d),

$$\overline{A_1} \cap A_2 = \varnothing, \quad A_1 \cap \overline{A_2} = \varnothing$$

In this case, A_1 and A_2 are said to be *separated*.

Example 1. (i) Let $X = (0, 2)$, $A = (0, 1)$ and $B = [1, 2)$ We can easily see that A and B are disjoint but not separated.

(ii) On the other hand if we take $B = (1, 2)$ and X and A same as above then A and B are separated sets.

Theorem 1. In order that the two sets A and B in a metric space (X, d) may be separated, it is sufficient but not necessary, that $d(A, B) > 0$.

Proof. Let $d(A, B) = k > 0$. Then the equation

$$k = \inf \{d(a, b): a \in A, b \in B\}$$

implies that $d(a, b) \geq k$ for every $a \in A$ and $b \in B$. If $a \in A$, then a cannot be a limit point of B, since the sphere $S_{k/2}(a)$ contains no point of B. Therefore $A \cap \overline{B} = \emptyset$. Similarly, $\overline{A} \cap B = \emptyset$. Hence A and B are separated sets.

The condition is not necessary, i.e. two sets can be at zero distance apart, and yet be separated. For example, the sets $\{x \in \mathbb{R} : x < 0\}$ and $\{x \in \mathbb{R} : x > 0\}$ are separated but at zero distance. On the other hand, the sets $\{x \in \mathbb{R} : x < 0\}$ and $\{x \in \mathbb{R} : x \geq 0\}$ are disjoint but not separated.

Theorem 2. Let (X, d) be a metric space and A, $B \subset X$ be separated. If $A_1 \subset A$ and $B_1 \subset B$, then A_1 and B_1 are separated.

Proof. Since $B_1 \subset B$, $\overline{B_1} \subset \overline{B}$. Hence, $A_1 \cap \overline{B_1} \subset A_1 \cap \overline{B} \subset A \cap \overline{B} = \emptyset$. Similarly, $\overline{A_1} \cap B_1 = \emptyset$. Hence, A_1 and B_1 are separated.

Theorem 3. Let (X, d) be a metric space. Then

(a) two closed sets in X are separated iff they are disjoint.
(b) two open sets in X are separated iff they are disjoint.

Proof

(a) Let F_1 and F_2 be two closed sets in X. Then this follows from the fact that $F_1 \cap \overline{F_2} = \overline{F_1} \cap F_2 = F_1 \cap F_2$, since $\overline{F_1} = F_1$ and $\overline{F_2} = F_2$.

(b) Let G_1 and G_2 be two open sets in X. We have to show that two disjoint open sets G_1 and G_2 are separated.
Let, if possible, G_1 and G_2 be not separated. Then, either $G_1 \cap \overline{G_2} \neq \emptyset$ or $\overline{G_1} \cap G_2 \neq \emptyset$. Suppose $G_1 \cap \overline{G_2} \neq \emptyset$ and $x \in G_1 \cap \overline{G_2}$. Then $x \in G_1$ and $x \in \overline{G_2}$. Therefore, for some $r > 0$,

$$S_r(x) \subset G_1 \text{ and } (S_r(x) - \{x\}) \cap G_2 \neq \emptyset$$

Hence, $G_1 \cap G_2$ is not empty, which is impossible since G_1 and G_2 are disjoint. Therefore, A and B are separated.

Theorem 4. Let (X, d) be a metric space. Now

(a) if the open set $G \subset X$ is the union of two separated sets A and B, then A and B are open.
(b) if the closed set $F \subset X$ is the union of two separated sets A and B, then A and B are closed.

Proof

(a) We may suppose A and B are not empty. For if $A = \emptyset$, then $B = G$, and A and B are both open. Let $x \in A$. Then $x \in G$. Since G is open, there exists a sphere $S_r(x) \subset G$, for some $r > 0$. Moreover

$$A \cap \overline{B} = \varnothing \text{ and } x \in A$$

imply that $x \notin \overline{B}$. Therefore, there exists a sphere $S_{r_1}(x)$, where $r_1 \leq r$, such that $S_{r_1}(x) \cap B = \varnothing$. Hence, $S_{r_1}(x) \subset A$, and hence $x \in A^0$. Thus every point of A is an interior point of A, since $x \in A$ was arbitrary. Hence A is open. Similarly, B is open.

(b) Since $F = A \cup B$ is closed, we have

$$A \cup B = \overline{A \cup B} = \overline{A} \cup \overline{B}$$

Hence

$$\overline{A} = \overline{A} \cap (\overline{A} \cup \overline{B}) = \overline{A} \cap (A \cup B) = (\overline{A} \cap A) \cup (\overline{A} \cup B)$$
$$= A \cup \varnothing = A,$$

i.e. A is closed. Similarly, we can show that B is closed.

6.2   Disconnected and Connected Sets

Definition 2. A metric space (X, d) is said to be *disconnected* if it is the union of two non-empty separated sets. X is said to be *connected* if it is not disconnected.

The subset A of a metric space X is said to be disconnected as a subspace of (X, d).

Theorem 5. A metric space X is connected iff the only nonempty subset of X which is both open and closed is X itself.

Proof. Let X be disconnected. Then X is the union of two separated sets, say B and C. Since X is closed, B and C both are closed. Since B and C are the complements of C and B, respectively, B and C are also open. Hence, a disconnected metric space is the union of two nonempty disjoint sets which are both open and closed.

If X is the only non-empty subset of X which is both open and closed, X is not the union of two non-empty disjoint sets which are both open and closed, and so X is not disconnected. Hence, X is connected.

Conversely, suppose that X is connected. If A is a non-empty subset which is both open and closed, A^c the complement of A is both closed and open. Thus, X is the union of two disjoint sets A and A^c which are both open and closed. Therefore, A and A^c are separated, which is impossible if A and A^c are both non-empty. But A is nonempty. Hence, $A^c = \varnothing$ and $A = X$, i.e. the only non-empty subset of X which is both open and closed is X.

Example 2. Every metric space X contains non-empty connected subsets. For each $x \in X$, the set $\{x\}$ is connected.

Note that the empty set is taken to be connected by definition. A finite set containing more than one point is not connected.

Example 3. The metric space $X = \mathbb{R} - \{0\}$ with the usual metric is disconnected because it is the union of two non-empty open sets $(-\infty, 0)$ and $(0, \infty)$.

Example 4. The set $\mathbb{Q}$ as a metric subspace of $\mathbb{R}$ is disconnected. For, if

$$A = \{x \in \mathbb{Q} : x < \sqrt{2}\} \text{ and } B = \{x \in \mathbb{Q} : x > \sqrt{2}\}$$

then $\mathbb{Q} = A \cup B$, where A and B both are of course open.

Last two examples are based on the following theorem.

Theorem 6. A metric space (X, d) is disconnected iff any one of the following is true:

  (i) X is the union of two nonempty disjoint open sets.
  (ii) X is the union of two nonempty disjoint closed sets.

Proof. Let X be disconnected. Then there exists a non-empty proper subset A of X which is both open and closed in X. Then A^c is also both open and closed in X and $X = A \cup A^c$. Hence, the set A and A^c satisfy the conditions.

Conversely, suppose that $X = A \cup B$ and $A \cap B = \varnothing$, where A and B are non-empty open sets. Hence, $A = X - B$ is closed also. Since B is non-empty, A is a proper subset of X. Thus A is a non-empty proper subset of X which is both open and closed. Hence, X is disconnected. We can use the similar arguments if A and B are closed, and obtain X to be disconnected.

Theorem 7. Let (X, d) be a metric space and $A \subset X$. Then A is disconnected iff there are two nonempty sets B and C such that: (i) $\overline{B}^X \cap C = \varnothing, B \cap \overline{C}^X = \varnothing$ and (ii) $A = B \cup C$, where $\overline{B}^X$ denotes the closure of B in X.

Proof. Let A be disconnected. Then the metric subspace A is disconnected in its own right, and so there exist nonempty $B, C \subset A$ such that

$$\overline{B}^A \cap C = \varnothing, B \cap \overline{C}^A = \varnothing \text{ and } A = B \cup C$$

Since $\overline{B}^A = \overline{B}^X \cap A$, it gives

$$\overline{B}^A \cap C = (\overline{B}^X \cap A) \cap C = \overline{B}^X \cap C$$

Hence $\qquad \overline{B}^X \cap C = \varnothing$. Similarly, $B \cap \overline{C}^X = \varnothing$.

Converse part follows immediately from the fact that $\overline{B}^X \supset \overline{B}^A$ and $\overline{C}^X \supset \overline{C}^A$, i.e. $\overline{B}^X \cap C = \varnothing$ implies $\overline{B}^A \cap C = \varnothing$. Similarly, $B \cap \overline{C}^A = \varnothing$.

6.3   Properties of Connected Sets

Theorem 8. The closure of a connected set is connected.

Proof. Let A be a connected subset of a metric space X. If $\overline{A}$ is disconnected, $\overline{A} = F_1 \cup F_2$, where F_1 and F_2 are disjoint nonempty closed sets. But

$$A = A \cap \overline{A} = (A \cap F_1) \cup (A \cap F_2)$$

expresses A as the union of two disjoint non-empty closed sets with respect to the connected subspace A, which is not possible. Hence $\overline{A}$ is connected.

Theorem 9. If A is a connected subset of a metric space X, and if $A \subset B \cup C$, where B and C are separated sets is X, then either $A \subset B$ or $A \subset C$.

Proof. Since B and C are separated sets and $A \cap B \subset B$, $A \cap C \subset C$ it follows (by Theorem 2) that $A \cap B$ and $A \cap C$ are separated sets. Also, we have

$$(A \cap B) \cup (A \cap C) = A \cap (B \cup C) = A$$

Thus, A has been expressed as the union of two nonempty separated sets. Therefore, A is disconnected which is a contradiction. Hence, either $A \subset B$ or $A \subset C$.

Theorem 10. If A is a connected subset of a metric space X, and if B is a subset of X such that $A \subset B \subset \overline{A}$, then B is connected. In particular $\overline{A}$ is connected.

Proof. Let, if possible, B be disconnected. Then there exists a separation of B, i.e. there exist sets F_1 and F_2 such that

$$F_1 \cap \overline{F_2} = \varnothing, \overline{F_1} \cap F_2 = \varnothing, \text{ and } F_1 \cup F_2 = B$$

Since $A \subset B$, we have $A \subset F_1 \cup F_2$.

Since the connected set A is contained in the union of two separated sets F_1 and F_2, it follows by Theorem 9 that $A \subset F_1$ or $A \subset F_2$.

Let $A \subset F_1$. Then $\overline{A} \subset \overline{F_1}$. Therefore

$$\overline{A} \cap F_2 \subset \overline{F_1} \cap F_2 = \varnothing$$

This implies that

$$\overline{A} \cap F_2 = \varnothing \qquad (1)$$

Further, since $B = F_1 \cup F_2$ and $B \subset \overline{A}$, $F_2 \subset B \subset \overline{A}$. Hence

$$\overline{A} \cap F_2 = F_2 \qquad (2)$$

From (1) and (2) we get $F_2 = \varnothing$ which contradicts that F_2 is non-empty. Hence, B is connected. Since $A \subset \overline{A} \subset \overline{A}$, it follows that $\overline{A}$ is connected.

Theorem 11. If two connected sets are not separated, their union is connected.

Proof. Let A and B be two connected sets which are not separated. If $A \cup B$ is not connected, then $A \cup B = C \cup D$, where C and D are non-empty separated sets. Then

$$A = A \cap (A \cup B) = A \cap (C \cup D) = (A \cap C) \cup (A \cap D)$$

which expresses A as the union of two separated sets, which is impossible since A is connected, unless one of $A \cap C$ and $A \cap D$ is empty. Suppose that $A \cap D$ is empty, so that $A \subset C$. Similarly, $B \subset C$ or $B \subset D$. If $B \subset C$, $A \cup B \subset C$, and hence D is empty which contradicts the hypothesis. Therefore, $B \subset D$. But this gives

$$A \cap \overline{B} \subset C \cap \overline{D} = \varnothing, \quad \overline{A} \cap B \subset \overline{C} \cap D = \varnothing$$

which is impossible since A and B are not separated. Hence, $A \cup B$ is connected.

Theorem 12. Let (X, d) be a metric space and let $\{A_\alpha : \alpha \in \wedge\}$ be a family of connected sets in X such that $\underset{\alpha \in \wedge}{\cap} A_\alpha \neq \varnothing$. Then $\underset{\alpha \in \wedge}{\cap} A_\alpha$ is connected.

Proof. Suppose on contrary that $\underset{\alpha \in \wedge}{\cap} A_\alpha$ is not connected. Then there are non-empty separated sets P and Q such that

$$\underset{\alpha \in \wedge}{\cap} A_\alpha = P \cup Q$$

Note that $A_\alpha \subset P$ and $A_\beta \subset Q$ for $\alpha, \beta \in \wedge$ $(\alpha \neq \beta)$, since P and Q are nonempty. Therefore, by Theorem 2, A_α and A_β are separated sets and $A_\alpha \cap A_\beta = \varnothing$ which is a contradiction to the hypothesis $\underset{\alpha \in \wedge}{\cap} A_\alpha \neq \varnothing$. Hence $\underset{\alpha \in \wedge}{\cup} A_\alpha$ is connected.

The following theorem is a generalization of the above theorem where the condition $\bigcap_{\alpha \in \wedge} A_\alpha \neq \varnothing$ is replaced by a weaker condition.

Theorem 13. Let (X, d) be a metric space and let $\{A_\alpha: \alpha \in \wedge\}$ be a family of connected sets in X such that there is a $A_{\alpha'}$ which is not separated from any of the remaining members of the family. Then $\bigcup_{\alpha \in \wedge} A_\alpha$ is connected.

Proof. Suppose that $\bigcup_{\alpha \in \wedge} A_\alpha$ is not connected. Let $B = \bigcup_{\alpha \in \wedge} A_\alpha$. Then we can write $B = C \cup D$, where C and D are non-empty separated sets, and so $A_{\alpha'} \subset C \cup D$. Now by Theorem 9, either $A_{\alpha'} \subset C$ or $A_{\alpha'} \subset D$. If $A_{\alpha'} \subset C$, then $A_{\alpha'} \cap D = \varnothing$. Since $A_{\alpha'} \cap A_\alpha$ is connected for each $\alpha \neq \alpha'$, we have

$$A_{\alpha'} \cup A_\alpha \subset C \cup D$$

Therefore, either $A_{\alpha'} \cup A_\alpha \subset C$ or $A_{\alpha'} \cup A_\alpha \subset D$. But $A_{\alpha'} \cup A_\alpha \subset D$ is not possible. Hence $A_{\alpha'} \cup A_\alpha \subset C$.

Therefore

$$\bigcup_{\alpha \in \wedge} (A_{\alpha'} \cup A_\alpha) \subset C$$

which implies that $B \subset C$ and $D = \varnothing$, which is a contradiction. Hence, $\bigcup_{\alpha \in \wedge} A_\alpha$ is connected.

Theorem 14. Let (X, d) be a metric space and $A \subset X$. If every pair of points in A lies in a connected subset of A, then A is connected.

Proof. Suppose A is disconnected. Then there exist disjoint open sets G_1 and G_2 of X such that $A \subset G_1 \cup G_2$ and $A \cap G_1$, $A \cap G_2$ are not empty. Let x be any point of $A \cap G_1$, y any point of $A \cap G_2$. By hypothesis, there is a connected subset B of A which contains both x and y. But $B \subset G_1 \cup G_2$, and $B \cap G_1$ and $B \cap G_2$ are not empty since they contain the points x and y, respectively. This is impossible since B is connected. Hence, A is not disconnected, i.e. A is connected.

6.4   Totally Disconnected Sets

In the Euclidean plane, every 'sphere' is a connected set. But it is not in general for every metric space. For example, the set $\mathbb{Q}$ of rationals with $d(x, y) = |x - y|$. The sphere $S_1(1)$ with centre 1 and radius 1 in $\mathbb{Q}$ is the set of all rationals x such that $0 < x < 2$. We can partition this set into two separated sets in many ways, e.g. we can take the subsets for which $0 < x^2 < 2$ and $0 < x^2 < 4$. Thus $S_1(1)$ is not connected.

Definition 3. Let (X, d) be a metric space and $A \subset X$. If every connected subset of A reduces to a single point, i.e. if no two points of A lie in a connected subset of A, we say that A is *totally disconnected*.

Example 5. In the usual metric space $\mathbb{R}$,

(a) $\mathbb{Q}$ is totally disconnected.
(b) the set of irrationals is totally disconnected.
(c) the set $(2, 3) \cup (3, 4)$ is disconnected but not totally disconnected.

Example 6. Another example can be obtained by taking any set and constructing a metric space with the discrete metric, i.e. any subset of a discrete space is totally disconnected.

6.5 Components

Let a and b be two points of a subset A of a metric space X. We write $a \sim b$ if there is a connected subset of A which contains both a and b. This relation connecting a pair of points of A is an equivalence relation.

The family of equivalence classes defined by this relation is called a *partition of A into components*. Each component is a connected set, and every point of A is a member of exactly one component. This component containing a is the union of all connected subsets of A which contain a, and so the largest connected set containing a. If A is connected set, it has only one component, A itself.

Definition 4. Let (X, d) be a metric space and $x \in X$. Then the *component* of X containing x is defined to be the largest connected subset of X containing x. We denote it by $C(x)$.

Since the metric space X is a subset of X, we can divide X itself into components.

Note that $C(x)$ is the union of all connected subsets of X each of which contains x.

Theorem 15. The components of a metric space are closed sets.

Proof. Let $C(x)$ be a component of X containing x. Since $C(x)$ is connected, so is $\overline{C(x)}$ (since $x \in C(x), x \in \overline{C(x)}$). Let $y \in \overline{C(x)}$. Then the connected set $\overline{C(x)}$ contains both x and y. But $C(x)$ is the largest connected set containing x, and thus $y \in C(x)$. Hence, $\overline{C(x)} \subset C(x)$, and $C(x)$ is closed.

Theorem 16. If a subset A of a metric space X is connected, open and closed, then A is a component of X.

Proof. Let $C(x)$ be a component of X containing x and $x \in A$. Then

$A \subset C(x)$. To show that A is a component of X, we have to show $A = C(x)$. Let $A \neq C(x)$. Then $A \cap C(x)$ is non-empty open set in $C(x)$ since A is open in X. Further, since A is closed in X, $X - A$ is open. Therefore, $C(x) \cap (X - A)$ is a non-empty open set in $C(x)$. Moreover, if $B = X - A$, we have

$$(C(x) \cap A) \cap (C(x) \cap B) = A \cap (C(x) - A) = \varnothing$$

and $$(C(x) \cap A) \cup (C(x) \cap B) = A \cup (C(x) - A) = C(x)$$

Therefore, $C(x)$ is the union of two non-empty disjoint open sets $C(x) \cap A$ and $C(x) \cap B$, and hence is disconnected. This contradiction leads to $A = C(x)$ and hence A is a component.

Theorem 17. Let (X, d) be a metric space and $x, y \in X$. Then either $C(x) = C(y)$ or $C(x) \cap C(y) = \varnothing$.

Proof. Let $y \in C(x)$. Then $C(x) \subset C(y)$, since $C(x)$ is a connected set containing y. If $x \in C(y)$, then by the same argument, we have $C(y) \subset C(x)$. Hence, $C(x) = C(y)$.

Now, let $y \notin C(x)$ and $C(x) \cap C(y) \neq \varnothing$. Let $z \in C(x) \cap C(y)$. Then we have $C(z) = C(x)$ and $C(z) = C(y)$ by first part. Hence, $C(x) = C(y)$ and so $y \in C(x)$ which is a contradiction. Therefore, $C(x) \cap C(y) = \varnothing$.

6.6 Connected Subsets of $\mathbb{R}$

Theorem 18. The usual metric space $\mathbb{R}$ is connected.

Proof. Let, if possible, $\mathbb{R}$ be disconnected. Then there exist non-empty closed sets G_1 and G_2 in $\mathbb{R}$ such that

$$\mathbb{R} = G_1 \cup G_2 \quad \text{and} \quad G_1 \cap G_2 = \varnothing$$

Choose a point $x \in G_1$ and a point $y \in G_2$. Since $G_1 \cap G_2 = \varnothing$, $x \neq y$. Let $x < y$. Then $[x, y] \subset \mathbb{R} = G_1 \cup G_2$. Hence, every point of $[x, y]$ is either in G_1 or in G_2.

Suppose that $u = \sup([x, y] \cap G_1)$ such that $u \in [x, y]$. Since u is the sup $([x, y] \cap G_1)$, for each $\varepsilon > 0$ there exists a $v \in [x, y] \cap G_1$ such that $u - \varepsilon < v < u$. This shows that every neighbourhood of u contains a point of $[x, y] \cap G_1$ and hence a point of G_1. Therefore, $u \in G_1$ or u is a limit point of G_1. Since G_1 is closed, we see that $u \in G_1$. Since $G_1 \cap G_2 = \varnothing$, $u \notin G_2$. Again since $y \in G_2$, we have $y \neq u$. This implies that $u < y$. Further, we have $u + \varepsilon \in G_2$ for every $\varepsilon > 0$ such that $u + \varepsilon < y$. This implies that every neighbourhood of u contains a point of G_2 other than u and hence it is a limit point of

G_2. Since G_2 is closed, $u \in G_2$. We have shown that $u \in G_1 \cap G_2$ which contradicts $G_1 \cap G_2 = \varnothing$. Hence, $\mathbb{R}$ is connected.

Theorem 19. A subset of $\mathbb{R}$ is connected iff it is an interval.

Proof. Let I be any non-empty interval of $\mathbb{R}$. Suppose that I is not connected and $I = A \cup B$, where A and B are two disjoint nonempty closed sets. Choose $a_1 \in A$ and $b_1 \in B$. Since $A \cap B = \varnothing$, $a_1 \neq b_1$. There is no loss of generality if we assume $a_1 < b_1$. We can bisect the interval (a_1, b_1). Then one of the two half-intervals of (a_1, b_1) has its left end point in A and right end point in B. For otherwise, if both the half-intervals have their left end points in A and right end points in B then A and B will have non-empty intersection which contradicts $A \cap B = \varnothing$. Let this interval, i.e. one-half of (a_1, b_1), be (a_2, b_2). Now bisect (a_2, b_2). Then one of the half-intervals of (a_2, b_2) will have its left end point in A and right end point in B. Let it be (a_3, b_3). Continuing this process, we get a nested sequence of intervals (a_n, b_n) with $a_n \in A$ and $b_n \in B$. These nested intervals have a common point, say, c which is precisely the common limit of the sequences $\{a_n\}$ and $\{b_n\}$. Since A and B are closed, c would be a common point of A and B. Hence a contradiction. Therefore, either A is empty or B is empty. Hence, I is connected.

Conversely, let I be connected and a and b be, respectively, the greatest lower bound and least upper bound. We have to show that I is an interval. Suppose the point p from the interval $a < p < b$ did not belong to I. Then $(-\infty, p)$ and (p, ∞) is a covering for I so that $I = (-\infty, p) \cup (p, \infty)$ with $a \in (-\infty, p)$ and $b \in (p, \infty)$. Since I is connected, one of these intervals will not intersect I. For example, suppose no point of I lies to the left of p which implies that p is the lower bound of I contradicting that a is the glb. Similarly, we can arrive at the contradiction that b is lub of I. Hence, $p \in (a, b)$ and I is the interval (a, b). The nature of the interval depends upon whether lub and glb are finite or infinite. So that I is one of the following:

$$(-\infty, b), (-\infty, b], (a, \infty), [a, \infty), (-\infty, \infty),$$

$$(a, b), [a, b), (a, b] \text{ and } [a, b]$$

6.7 Relation Between Connectedness and Continuity

Theorem 20. Let f be a continuous function from a metric space (X, d_1) into a metric space (Y, d_2). If X is connected, then $f(X)$ is connected in Y, i.e. the continuous image of a connected set is connected.

Proof. Let $A = f(X)$ so that $f : X \to A$. Let, if possible A be disconnected.

Then there exists a non-empty proper subset B of A such that B is both open and closed in A. But since f is continuous from $X \to f(X) \subset Y, f^{-1}(B)$ would be nonempty proper subset of X i.e. both open and closed in X. This contradicts the hypothesis that X is connected. Therefore $f(X)$ is not disconnected, i.e. $f(X)$ is connected.

Corollary 21. Let (X, d) be a connected metric space and $f : X \to \mathbb{R}$ be continuous. Then, the range $f(X)$ in an interval.

Proof. By Theorem 20, $f(X)$ is a connected subset of $\mathbb{R}$. Therefore, by Theorem 19, $f(X)$ is an interval.

Corollary 22 (Intermediate value theorem). If f is a continuous real valued function on the closed and bounded interval $[a, b]$, then f takes every value between $f(a)$ and $f(b)$.

Proof. Let f be real valued continuous function on the bounded closed interval $I = [a, b]$. Then by Theorem 19, $[a, b]$ is a connected subset of $\mathbb{R}$. Since f is continuous, $f([a, b])$ is connected subset of $\mathbb{R}$. Hence $f([a, b])$ is an interval, (by Theorem 19). Therefore, if $f(a) < c < f(b)$, there exists a point $x \in [a, b]$ such that $f(x) = c$.

Corollary 23. A metric space (X, d) is disconnected iff there exists a continuous function from X onto the discrete two point space $\{0, 1\}$.

Proof. Let X be disconnected. Then $X = A \cup B$, where A and B are nonempty, disjoint, open subsets of X. Define $F : X \to \{0, 1\}$ by

$$f(x) = \begin{cases} 0, & x \in A, \\ 1, & x \in B; \end{cases}$$

which is clearly a continuous function from X onto $\{0, 1\}$.

Conversely, let there exist a continuous function from X onto $\{0, 1\}$. Suppose, if possible, X is connected. Then by Theorem 20, the space $\{0, 1\}$ is connected. Hence, a contradiction. Thus X is disconnected.

Theorem 24. If f is continuous real valued function on the compact connected metric space X, then f takes on every value between its minimum value and maximum value.

Proof. Let $f : X \to \mathbb{R}$ be continuous. There are two cases:

(i) Let f be a constant function. Then the range of f is a single point. Let $f(x) = c$ for every $x \in X$. Then the maximum and minimum values of f coincide with c.

(ii) Let f be a non-constant function. Then the range of f contains more than one point. Since X is connected, $f(X)$ is connected subset of $\mathbb{R}$ and hence $f(X)$ is an interval of $\mathbb{R}$. Since X is compact, $f(X)$ is a closed and bounded interval of $\mathbb{R}$. Further, since X is compact and f is continuous real valued on X, f takes the maximum and minimum values at points of X. Let a and b be points of X such that $f(a) = \underset{x \in X}{\text{glb}} f(x)$ and $f(b) = \underset{x \in X}{\text{lub}} f(x)$. Since $f(X)$ is a bounded and closed interval, it should be $[f(a), f(b)]$. Hence, f takes every value between its minimum and maximum values.

6.8 Miscellaneous Examples

1. A compact set need not be connected and a connected set need not be compact.
   Consider the set $A = \{0, 1, 1/2, 1/3, \ldots\}$. A is closed and bounded set in $\mathbb{R}$ so that it is compact but it is not connected. On the other hand the interval (a, b) is connected but not compact.
2. A subset of a connected set which is not connected: $\mathbb{R}$ is connected in the usual metric but its subset $\{1, 2, 3, \ldots\}$ is not connected.
3. If A and C are connected subsets of a metric space and if $A \subset B \subset C$, then B is not necessarily connected.
   Let $A = (0, 1)$, $B = (0, 1) \cup [1, 2]$, $C = [0, \infty)$. Then $A \subset B \subset C$. We see that A and C are connected but B is not connected.
4. The interval $[0, 1]$ is not a connected subset of $\mathbb{R}_d$.
   Let $x \in [0, 1]$ Since each singleton set is both open and closed in $[0, 1]$. $\{x\}$ is both open and closed in $[0, 1]$. Hence, $[0, 1]$ is not a connected set of $\mathbb{R}_d$.
5. Let E be a subset of a metric space X. Then E is connected if every two points of E are contained in some connected subset of E.
   Suppose E is not connected. Then there exist two non-empty subsets A and B of X such that $E = A \cup B$, $A \cap \overline{B} = \varnothing$, $\overline{A} \cap B = \varnothing$. Since A and B are non-empty, there exist a point $a \in A$ and a point $b \in B$. Therefore, a and b must be contained in some connected subset F of E. By Theorem 9, $F \subset A$ or $F \subset B$, since $F \subset A \cup B$. Then it follows that either $a, b \in A$ or $a, b \in B$ which is a contradiction, since $a \in A$ and $b \in B$, $A \cap B = \varnothing$. Hence, E must be connected.

Hence, f is a contraction on $[0, 1]$.

The following theorem gives a relation between contraction and continuity.

Theorem 1. Let (X, d) be a metric space and T be a contraction on X. Then T is continuous on X.

Proof. Let $\varepsilon > 0$ and $x_0 \in X$. Then from the definition of contraction

$$d(Tx, Tx_0) \leq \alpha \, d(x, x_0), \, 0 \leq \alpha < 1$$

If $d(x, x_0) < \varepsilon$, then

$$d(Tx, Tx_0) < \alpha \, \varepsilon < \varepsilon$$

Hence T is continuous on X.

7.2 Banach's Fixed Point Theorem

Definition 2. Let (X, d) be a metric space and $T : X \to X$ be a mapping. The point $x \in X$ is called a *fixed point* of T if $Tx = x$.

Example 3

   (i) The mapping $x \to x + c$ of $\mathbb{R}$ into itself has no fixed point, where c is a real constant.

   (ii) The mapping $x \to x^2$ of $\mathbb{R}$ into itself has two fixed points 0 and 1.

   (iii) The mapping $x \to x$ of $\mathbb{R}$ into itself has infinitely many fixed points.

Theorem 2 (Banach's fixed point). Let (X, d) be a complete metric space and let $T : X \to X$ be contraction on X. Then, T has a unique fixed point in X.

Proof. Since T is a contraction on X, we get

$$d(Tx, Ty) \leq \alpha \, d(x, y) \tag{1}$$

for $0 \leq \alpha < 1$ and for all $x, y \in X$.

By further application of the above relation, we get

$$d(T^2x, T^2y) \leq \alpha \, d(Tx, Ty) \leq \alpha^2 \, d(x, y)$$

Proceeding in this manner, we can establish the following relation for any n

$$d(T^nx, T^ny) \leq \alpha^n d(x, y) \tag{2}$$

For any point $x_0 \in X$, we can construct a sequence $\{x_n\}$ of points in X as

$$x_1 = Tx_0, \; x_2 = Tx_1 = T^2x_0, \ldots, x_n = Tx_{n-1} = T^nx_0$$

We first show that $\{x_n\}$ is a Cauchy sequence in X. If m, n are positive integers and $m > n$, let $m = n + p$. By using the triangle inequality successively, we get

$$d\,(x_n, x_m) = d(x_n, x_n + p)$$

$$\leq d(x_n, x_{n+1}) + d(x_{n+1}, x_{n+2}) + \ldots + d\,(x_{n+p-1}, x_{n+p})$$

$$= d(T^nx_0, T^nx_1) + d(T^{n+1}x_0, T^{n+1}x_1)$$

$$+ \ldots + d(T^{n+p-1}x_0, T^{n+p-1}x_1)$$

By using (2), we obtain

$$d(x_n, x_m) \leq \alpha^n d(x_0, x_1) + \alpha^{n+1}d(x_0, x_1) + \ldots + \alpha^{n+p-1}d(x_0, x_1)$$

$$\leq \alpha^n d(x_0, x_1)\,[1 + \alpha + \alpha^2 + \ldots]$$

$$\leq \frac{\alpha^n}{1 - \alpha}d(x_0, x_1), 0 \leq \alpha < 1$$

$$\rightarrow 0, \text{ as } n \rightarrow \infty$$

Hence $\{x_n\}$ is a Cauchy sequence,

Since X is complete, there exists $x \in X$ such that $x_n \rightarrow x(n \rightarrow \infty)$. We shall show that $Tx = x$.

Since T is continuous, we have $Tx = \lim_n Tx_n = \lim_n x_{n+1} = x$, i.e. x is a fixed point of T.

Now we show that this fixed point is unique. Let, if possible, x and y be two fixed points of T in X. Then $Tx = x$ and $Ty = y$. We have

$$d(x, y) = d(Tx, Ty) \leq \alpha\, d(x, y)$$

This gives $1 \leq \alpha$ which is a contradiction to $0 \leq \alpha < 1$. Hence $x = y$, i.e. x is a unique fixed point of T.

Remark 1. The condition of completeness from the hypothesis of the above theorem cannot be dropped as shown by the following example.

Example 3. Let us take $X = (0, 1]$ with the absolute value metric. Clearly X is an incomplete metric space.

Let $T : X \rightarrow X$ be defined by $Tx = \frac{1}{2}x$. Then 0 is the fixed point of the mapping T, but $0 \notin X$. Hence, T has no fixed point.

Remark 2. When $\alpha = 1$, the above theorem may fail to be true.

Example 4. Define $T : \mathbb{R}^+ \rightarrow \mathbb{R}^+$ by $Tx = (1 + x^2)^{1/2}$. T has no fixed point. Let, if possible, T have a fixed point then $Tx = x$. This implies $(1 + x^2)^{1/2} = x$ which is not possible.

Theorem 3. Let (X, d) be a complete metric space and $T : X \rightarrow X$ be a mapping. If T^n is a contraction on X for some $n \in \mathbb{N}$, then T has a unique fixed point.

Proof. Let $S = T^n$. By Theorem 2, S has a unique fixed point x_0. Let us show that x_0 is also a fixed point of T.

Since $S = T^n$, it follows that $TSx = STx$ for all x in X. Since S is a contraction, there is an α, $0 \le \alpha < 1$ such that

$$d(Sx, Sy) \le \alpha \, d(x, y)$$

for all x and y in X. If $Tx_0 \ne x_0$ we get the contradiction

$$0 < d(Tx_0, x_0) = d(TSx_0, Sx_0)$$

$$= d(STx_0, Sx_0) \le \alpha d(Ix_0, x_0) < d(Tx_0, x_0)$$

Hence $Tx_0 = x_0$. Also, since every fixed point of T is a fixed point of S, it follows that T has a unique fixed point.

7.3 Applications

This section gives some of the applications of the Banach's fixed point theorem.

1. Finite system of Linear Equations

Consider a system of n linear equations in n unknowns

$$y_i = \sum_{j=1}^{n} a_{ij} x_j + b_i \, (i = 1, 2, \ldots, n) \tag{1}$$

The constants a_{ij} and b_i are complex numbers.

Let us consider the space $\mathbb{R}^n$ with the metric

$$d(x, y) = \max_i |x_i - y_i|$$

On $\mathbb{R}^n$, let us define $T : \mathbb{R}^n \rightarrow \mathbb{R}^n$ by

$$y = Tx = Ax + B \tag{2}$$

System (1) can be written in the form of (2), where $A = (a_{ij})$ is $n \times n$ matrix, $x = [x_1, x_2, \ldots, x_n]$ and $B = [b_1, b_2, \ldots, b_n]$ are column vectors.

where f is a real valued, continuous function defined on $\mathbb{R} \times \mathbb{R}$. We shall seek a solution $y(t)$ for (7) which satisfies the initial condition

$$y(t_0) = y_0$$

This is said to be the initial value problem. Since f is continuous, it is easily checked that a solution of the initial value problem is equivalent to a solution of the integral equation

$$y(t) = y_0 + \int_{t_0}^{t} f(y(s), s)ds \tag{8}$$

Let us consider the operator $z = F(y)$ where

$$z(t) = y_0 + \int_{t_0}^{t} f(y(s), s)\,ds$$

We observe that if y is continuous, then z is continuous. Thus $F : C(I) \to C(I)$, where $C(I)$ is the space of real valued continuous functions defined on some interval I containing t_0.

Now $y(t)$ is a solution of (8) iff $y = F(y)$; i.e. iff y is a fixed point of F. Now we have to find under what condition (on f) is the mapping F a contraction.

For our convenience we write $y_0 = t_0 = 0$.

Since f is continuous, it is bounded on the set $-1 \le y \le 1$, $-1 \le t \le 1$, i.e.

$$|f(y, t)| \le M$$

on this set, Now assume that f satisfies

$$|f(y, t) - f(x, t)| \le K\,|y - x|$$

for all t, x, y satisfying $-1 \le t \le 1$, $-1 \le x \le 1$, $-1, \le y \le 1$.

With M and K defined above, let X denote the set of all continuous real valued functions $\Phi(t)$ satisfying

$$|\Phi(t)| \le M\,|t|$$

on the interval $I = [-T, T]$, where $0 < T \le 1$, $MT \le 1$ and $KT < 1$. X is then a subset of $C(I)$. Moreover, if $C(I)$ has the sup-metric d_∞, then X is a closed subset. To prove this, we want to show that the complement is open. Hence, let Φ be in X^C. Then for some $u \in I$, $|\Phi(u)| > M|u|$. Let $2\varepsilon = |\Phi(u)| - M|u| > 0$. Now if Ψ is in $S_\varepsilon(\Phi)$, then $|\Psi(u)| - M|u| > \varepsilon$. Hence $\Psi \in X^C$, X^C thus is open.

Since $(C(I), d_\infty)$ is complete and X is a closed subspace, if follows that (X, d_∞) is complete. Let us now show that F maps X into X, and that F is a contraction.

Let $\Phi \in X$, then $F(\Phi)$ is in X since

$$|F(\Phi)(t)| \le \int_0^{|t|} |f(\Phi(s), s)| ds \le \int_0^{|t|} M ds \le M|t|,$$

for all $t \in I$. Let x, y be in X. Then

$$F(x)(t) - F(y)(t) = \int_0^t [f(x(s), s) - f(y(s), s)] ds$$

Thus for $t \ge 0$ we get

$$|F(x)(t) - F(y)(t)| \le \int_0^t K|x(s) - y(s)| ds$$

$$\le \int_0^t K d_\infty(x, y) ds \le Kt\, d_\infty(x, y)$$

For $t \le 0$, we get $|F(x)(t) - F(y)(t)| \le K|t| d_\infty(x, y)$.
For $|t| \le T$, we get

$$d_\infty(F(x), F(y)) \le KT\, d_\infty(x, y) = \alpha\, d_\infty(x, y)$$

Since $\alpha = KT < 1$, F is a contraction. Therefore, by Banach fixed point theorem F has a unique fixed point and the initial value problem has a unique solution on $[-T, T]$.

4. Integral Equations
Consider the integral equation

$$x(t) = \lambda \int_a^t K(t, u) x(u) du + \Phi(t) \tag{9}$$

where λ is an arbitrary parameter, Φ is continuous on $[a, b]$ and K is continuous on the triangle $a \le u \le t \le b$.

We shall see that the integral equation (9) which is known as the Volterra's integral equation, has a unique solution (continuous) for $x = x(t)$ on $[a, b]$ for every λ.

Let $X = C[a, b]$ be a complete metric space of all continuous functions $x(t)$ with the metric

$$d(x, y) = \max_{a \le t \le b} |x(t) - y(t)|$$

Let us consider $T : C[a, b] \to C[a, b]$ defined by

$$Tx(t) = \lambda \int_a^t K(t, u) x(u)\, du + \Phi(t) \tag{10}$$

This is well defined, since x continuous on $[a, b]$ implies that the integral exists and is a continuous function of the upper limit. Since Φ and K are continuous, they are bounded. Let $| \Phi(t) | \leq m$, $| K(t, u) | \leq M$. Then

$$| Tx(t) - Ty(t) | = \left| \lambda \int_a^t K(t, u)[x(u) - y(u)]\, du \right|$$

$$\leq | \lambda | M(t - a)\, d(x, y)$$

Thus, T is a contraction only for sufficiently small values of $| \lambda |$.

Now for $x, y \in C[a, b]$ an easy induction shows that

$$| T^n x(t) - T^n y(t) | \leq | \lambda |^n M^n \frac{(t - a)^n}{n!} d(x, y) \tag{11}$$

Therefore

$$d(T^n x, T^n y) \leq \frac{1}{n!} | \lambda |^n M^n (b - a)^n d(x, y)$$

But

$$| \lambda |^n M^n (b - a)^n / n! \to 0 \text{ as } n \to \infty$$

Hence, choose n so that

$$\alpha = \frac{| \lambda |^n M^n (b - a)^n}{n!}$$

lies between 0 and 1.

Hence with such a choice of n

$$d(T^n x, T^n y) \leq \alpha d(x, y)$$

and thus the mapping T^n is a contraction on $C[a, b]$ for any choice of λ. Therefore, by Theorem 3, $x = Tx$ has a unique solution for x, i.e. (9) has a unique solution $x \in C[a, b]$.

7.4 Measures of Noncompactness

The first measure of noncompactness, function α, was defined and studied by Kuratowski [6]. Later Darbo [3] used the function α in fixed point theory. We will also discuss other measures of noncompactness, e.g. Hausdorff measure and Istrătesku measure. Measures of noncompactness have proved useful in several areas of functional analysis, operator theory and differential equations (see for example [4]). Most recently, Malkowsky and Rakočević [8] have used measures of noncompactness to the study of matrix transformations.

1. Kuratowski Measure

Definition 3. Let (X, d) be a metric space and Q a bounded subset of X. Then the *Kuratowski measure of noncompactness* of Q, denoted by $\alpha(Q)$, is the infimum of the set of all numbers $\varepsilon > 0$ such that Q can be covered by a finite number of sets with diameters less than ε, i.e.

$$\alpha(Q) := \inf \left\{ \begin{array}{l} \varepsilon > 0 : Q \subset \bigcup_{i=1}^{n} S_i, \ S_i \subset X, \\ \operatorname{diam}(S_i) < \varepsilon \, (i = 1, 2, \ldots, n; n \in \mathbb{N}) \end{array} \right\} \tag{1}$$

The function α is called *Kuratowski's measure of noncompactness* or *α-measure* or *setmeasure*. Clearly

$$\alpha(Q) \leq \operatorname{diam}(Q) \text{ for each bounded subset } Q \text{ of } X \tag{2}$$

As an immediate consequence, we obtain

Theorem 4. Let Q, Q_1 and Q_2 be bounded subsets of a complete metric space (X, d). Then

(a) $\alpha(Q) = 0$ iff $\overline{Q}$ is compact

(b) $\alpha(Q) = \alpha(\overline{Q})$

(c) $Q_1 \subset Q_2$ implies $\alpha(Q_1) \leq \alpha(Q_2)$

(d) $\alpha(Q_1 \cup Q_2) = \max \{\alpha(Q_1), \alpha(Q_2)\}$

(e) $\alpha(Q_1 \cap Q_2) \leq \min \{\alpha(Q_1), \alpha(Q_2)\}$

Proof. The statements (a) and (c) follow directly from Definition 3. Clearly $\alpha(Q) \leq \alpha(\overline{Q})$. Let $\varepsilon > 0$, S_i be a bounded subset of X with diam $(S_i) < \varepsilon$ for $i = 1, 2, \ldots, n$, and $Q \subset \bigcup_{i=1}^{n} S_i$. Then $\overline{Q} \subset \bigcup_{i=1}^{n} \overline{S}_i$. Since $\operatorname{diam}(S_i) = \operatorname{diam}(\overline{S}_i)$, we conclude $\alpha(\overline{Q}) \leq \alpha(Q)$. This proves (b).

From (c), we have $\alpha(Q_1) \leq \alpha(Q_1 \cup Q_2)$ and $\alpha(Q_2) \leq \alpha(Q_1 \cup Q_2)$, and so

$$\max \{\alpha(Q_1), \alpha(Q_2)\} \leq \alpha(Q_1 \cup Q_2) \tag{3}$$

Let max $\{\alpha(Q_1), \alpha(Q_2)\} = s$ and $\varepsilon > 0$. By Definition 3, we know that Q_1 and Q_2 can be covered by a finite number of subsets of diameter smaller than $s + \varepsilon$. Obviously, the union of these covers is a finite cover of $Q_1 \cup Q_2$. Hence, we have $\alpha(Q_1 \cup Q_2) \leq s + \varepsilon$, and now we get (d) from (3). From $Q_1 \cap Q_2 \subset Q_1$ and $Q_1 \cap Q_2 \subset Q_2$ we obtain $\alpha(Q_1 \cap Q_2) \leq \alpha(Q_1)$ and $\alpha(Q_1 \cap Q_2) \leq \alpha(Q_2)$. Hence, $\alpha(Q_1 \cap Q_2) \leq \min \{\alpha(Q_;), \alpha(Q_2)\}$. This gives (e).

2. Generalized Cantor Intersection Theorem

The next theorem is a generalization of the well-known Cantor Intersection theorem. This is due to Kuratowski.

Theorem 5. Let (X, d) be a complete metric space. If $\{F_n\}$ is a decreasing sequence of non-empty, closed and bounded subsets of X such that $\alpha(F_n) \to 0 (n \to \infty)$, then the intersection $F_\infty = \overset{\infty}{\underset{n=1}{\cap}} F_n$ is a non-empty and compact subset of X.

Proof. The set F_∞ is a closed subset of X, since each F_n is a closed subset of X. Since $F_\infty \subset F_n$ for all $n = 12, 3 \ldots$, we obtain from (a) and (c) of Theorem 4 that F_∞ is a compact set. Now we have to show that F_∞ is non-empty. Let $x_n \in F_n (n = 1, 2, \ldots)$ and $X_n = \{x_i : i \geq n\}$ for $n = 1, 2, \ldots$

Since $X_n \subset F_n$, we obtain from (a), (c) and (d) of Theorem 4,

$$\alpha(X_1) = \alpha(X_n) \leq \alpha(F_n) \text{ for each } n \tag{4}$$

Since $\alpha(F_n) \to 0 (n \to \infty)$, we obtain $\alpha(X_1) = 0$ and hence X_1 is a relatively compact set. Thus the sequence $\{x_n\}$ has a convergent subsequence with limit $x \in X$, say. Since F_n is closed in X, we obtain $x \in F_n$ for all $n = 1, 2, \ldots$, i.e. $x \in F_\infty$. Therefore, $F_\infty \neq \varnothing$.

3. Hausdorff Measure

Definition 4. Let (X, d) be a metric space and Q a bounded subset of X. Then the *Hausdorff measure of noncompactness* of the set Q, denoted by $\lambda(Q)$ is defined to be the infimum of the set of all reals $\varepsilon > 0$ such that Q can be covered by a finite number of open balls of radii less than ε, i.e.

$$\lambda(Q) = \inf \left\{ \begin{array}{l} \varepsilon > 0 : Q \subset \overset{n}{\underset{i=1}{\cup}} B(x_i, r_i), x_i \in X, \\ r_i < \varepsilon (i = 1, 2, \ldots, n) n \in \mathbb{N} \end{array} \right\} \tag{5}$$

The function λ is called *Hausdorff measure of noncompactness* or *ball measure of noncompactness* or *λ-measure*.

Note that in this definition it is not supposed that centres of the balls which cover Q belong to Q. Hence (5) can equivalently be stated as follows:

$$\lambda(Q) = \inf \{\varepsilon > 0: Q \text{ has a finite } \varepsilon\text{-net in } X\} \tag{6}$$

The next theorem can be proved analogously as in the case of Kuratowski measure of noncompactness.

Theorem 6. Let Q, Q_1 and Q_2 be bounded subsets of the metric space (X, d). Then

(a) $\lambda(Q) = 0$ iff Q is totally bounded

(b) $\lambda(Q) = \lambda(\overline{Q})$

(c) $Q_1 \subset Q_2$ implies $\lambda(Q_1) \leq \lambda(Q_2)$

(d) $\lambda(Q_1 \cap Q_2) = \max\{\lambda(Q_1), \lambda(Q_2)\}$

(e) $\lambda(Q_1 \cap Q_2) \leq \min\{\lambda(Q_1), \lambda(Q_2)\}$

Theorem 7. Let (X, d) be a metric space and Q be a bounded subset of X. Then

$$\lambda(Q) \leq \alpha(Q) \leq 2\lambda(Q) \tag{7}$$

Proof. Let $\varepsilon > 0$. If $\{x_1, x_2, \ldots, x_n\}$ is an ε-net of Q, then $\{Q \cap B(x_i, \varepsilon)\}_{i=1}^n$ is a cover of Q with sets of diameter less than 2ε. This gives $\alpha(Q) \leq 2\lambda(Q)$. Now suppose that $\{S_i\}_{i=1}^n$ is a cover of Q with sets of diameter less than ε and $y_i \in S_i$ for $i = 1, 2, \ldots, k$. Now $\{y_1, \ldots, y_k\}$ is an ε-net of Q. Therefore, $\lambda(Q) \leq \alpha(Q)$. Combining these two we get (7).

4. Hausdorff Distance

Definition 5. Let M_X be the set of all nonempty and bounded subsets of a metric space (X, d), and let M_X^C be the subfamily of M_X consisting of all closed sets. Further, let N_X be the set of all nonempty and relatively compact subsets of (X, d). Let $d_H : M_X \times M_X \to \mathbb{R}$ be the function defined by

$$d_H(S,Q) = \max\left[\sup_{x \in S} d(x, Q), \sup_{y \in Q} d(y, S)\right], (S, Q \in M_X) \tag{8}$$

The function d_H is called *Hausdorff distance* and $d_H(S, Q)$ $(S, Q \in M_X)$ is the *Hausdorff distance of sets S and Q*.

Theorem 8. Let (X, d) be a metric space. Then (M_X^C, d_H) is a metric space.

Proof. We see that $d_H(S, Q) = 0$ iff $S = Q$, and also

$$d_H(S, Q) = d_H(Q, S) \text{ for all } S, Q \in M_X^C$$

For triangle inequality, suppose that $S, Q, F \in M_X^C, x \in S, y \in Q$ and $z \in F$. It is easy to prove that $d(x, F) \leq d(x, y) + d(y, F) \leq d(x, y) + d_H(Q, F)$ and this implies

$$d(x, F) \leq \inf_{y \in Q} d(x, y) + d_H(Q, F) = d(x, Q) + d_H(Q, F)$$

$$\leq d_H(S, Q) + d_H(Q, F) \tag{9}$$

Replacing x and F by z and S in (9), respectively, we have

$$d(z, S) \leq d_H(F, Q) + d_H(Q, S) \tag{10}$$

Thus, (9) and (10) together imply that $d_H(S, F) \leq d_H(S, Q) + d_H(Q, F)$, and so (M_X^C, d_H) is a metric space.

The following theorem gives a relation between Hausdorff measure and Hausdorff distance.

Theorem 9. Let (X, d) be a metric space, $Q, Q_1, Q_2 \in M_X$, and N_X^C be the set of all nonempty and compact subsets of (X, d). Then

$$|\lambda(Q_1) - \lambda(Q_2)| \leq d_H(Q_1, Q_2) \tag{11}$$

$$\lambda(Q) = d_H(Q, N_X^C) \tag{12}$$

Proof. Let $\varepsilon > 0$ and $d = d_H(Q_1, Q_2)$. Then from (8) and (5), it follows that there is a finite set $S \subset X$ such that

$$Q_1 \subset B(Q_2, d + \varepsilon) \text{ and } Q_2 \subset B(S, \lambda(Q_2) + \varepsilon) \tag{13}$$

Furthermore, (13) implies

$$Q_1 \subset B(S, d + \lambda(Q_2) + 2\varepsilon) \tag{14}$$

and hence $\qquad \lambda(Q_1) \leq \lambda(Q_2) + d + 2\varepsilon$

Now (15) implies (11). For (12), we note that

$$\lambda(Q) \leq d_H(Q, N_X^C) \tag{16}$$

follows from (11). Now for reverse inequality, let $\varepsilon > 0$ be given. Then there exists a finite set $F \subset X$, such that

$$Q \subset B(F, \lambda(Q) + \varepsilon) \text{ and } F \subset B(Q, \lambda(Q) + \varepsilon) \tag{17}$$

Now (17) and (8) together imply

$$d_H(Q, N_X^C) \le d_H(Q, F) \le \lambda(Q) + \varepsilon \qquad (18)$$

Since ε was arbitrary, (16) and (18) give (12).

5. Inner Hausdorff Measure

Definition 6. Let (X, d) be a metric space and Q a bounded subset of X. Then the *inner Hausdorff measure of noncompactness* of the set Q, denoted by $\lambda_*(Q)$ is defined to be the infimum of the set of all positive reals ε such that Q can be covered by a finite number of balls of radii less than ε with centres in Q, i.e.

$$\lambda_*(Q) = \inf \left\{ \begin{array}{l} \varepsilon > 0: Q \subset \overset{n}{\underset{i=1}{\cup}} B(x_i, r_i), x_i \in Q, \\ r_i < \varepsilon \, (i, = 1, 2, \ldots, n) \, n \in \mathbb{N} \end{array} \right\} \qquad (19)$$

The function λ_* is called *inner Hausdorff measure of noncompactness*. Equivalently we can state that

$$\lambda_*(Q) = \inf\{\varepsilon > 0: Q \text{ has a finite } \varepsilon\text{-net in } Q\} \qquad (20)$$

Note that if the centres of the balls in Definition 4 are in Q then we have Definition 6. The following theorem has some distinction from Theorems 4 and 6.

Theorem 10. Let Q, Q_1 and Q_2 be bounded subsets of the metric space (X, d). Then

$$\lambda_*(Q) = 0 \text{ iff } Q \text{ is totaly bounded} \qquad (21)$$

$$\lambda_*(Q) = \lambda_*(\overline{Q}) \qquad (22)$$

but in general

$$Q_1 \subset Q_2 \text{ does not imply } \lambda_*(Q_1) \le \lambda_*(Q_2) \qquad (23)$$

and $\qquad \lambda_*(Q_1 \cup Q_2) \ne \max \{\lambda_*(Q_1), \lambda_*(Q_2)\} \qquad (24)$

6. Istrătesku Measure

The following measure of noncompactness is due to Istrătesku [4] which is closely related to the Hausdorff and Kuratowski measures.

Definition 7. A bounded subset Q of a complete metric space (X, d) is said to be ε-*discrete* if $d(x, y) \ge \varepsilon$ for all $x, y \in Q$ with $x \ne y$.

Definition 2. A subset Y of a linear space X is said to be a *linear subspace* if $x_1 + x_2 \in Y$ whenever $x_1, x_2 \in Y$ and $\alpha x \in Y$ whenever $\alpha \in \mathbf{K}$ and $x \in Y$.
Note that a linear subspace is itself a linear space.

Example 1. (i) $\mathbb{R}^n$ is a real linear space if we define coordinatewise operations as follows:

$$x + y = (x_1 + y_1, \ldots, x_n + y_n), \ \lambda x = (\lambda x_1, \ldots, \lambda x_n)$$

$$\theta = (0, 0, \ldots, 0) \text{ and } -x = (-x_1 - x_2, \ldots, -x_n)$$

Similarly, we can define such operations for $\mathbb{C}^n$ to be a linear space.

(ii) w is a linear space under the operations

$$\{x_n\} + \{y_n\} = \{x_n + y_n\}, \ \lambda\{x_n\} = \{\lambda x_n\}$$

(iii) The sequence spaces c, c_0, l_∞ and l_P ($1 \le p < \infty$) are all linear with the coordinatewise operations as defined in w. Moreover, these are linear subspaces of w.

(iv) The space $C[0, 1]$ is linear with the linear operations defined by

$$(\lambda f_1 + \mu f_2)(x) = \lambda f_1(x) + \mu f_2(x)$$

Note that $\lambda f_1 + \mu f_2 \in C[0, 1]$ whenever $f_1, f_2 \in C[0, 1]$.

(v) The set of all $n \times m$ matrices can be viewed as a linear space.

8.2   Linear Transformations

There is a special class of transformations that plays a central role in the case of linear spaces, namely, linear transformations.

Definition 3. A transformation L of a linear space X into a linear space Y, where X and Y have the same scalar field is said to be a *linear transformation* if

$$L(\alpha x + \beta y) = \alpha L(x) + \beta L(y) \text{ for all } x, y \in X \text{ and all } \alpha, \beta \in \mathbf{K}$$

Example 2. (i) The zero mapping $0 : X \to Y$ defined by $0(x) = \theta$ for all $x \in X$, is a linear transformation.

(ii) The identity map $i : X \to X$ defined by $i(x) = x$ is a linear transformation.

(iii) A constant map $f : \mathbb{R} \to \mathbb{R}$ defined by $f(x) = a$ (where $a \in \mathbb{R}$ is fixed) is not a linear transformation.

Example 3. Suppose that we are interested in the temperature of an infinite bar as a function of time t, and position x. Denote the temperature by $T(x, t)$. Further, assume that bar is being heated along its length by a distributed heat source.

The heat supplied per unit length at x and t is denoted by $H(x, t)$. Assume that at $t = 0$, $T(x, 0) = 0$. Assume that H is a point in the linear space X made up of all bounded continuous, real-valued functions defined on $(-\infty, \infty) \times [0, \infty)$. It is well-known that

$$T(x, t) = \int_0^\infty \int_{-\infty}^\infty H(x - x', t - t') H(x', t') dx' dt' \qquad (1)$$

where
$$H(x, t) = \begin{cases} M \dfrac{e^{-x^2/4t}}{\sqrt{t}}, & t > 0, \\ 0, & t \leq 0, \end{cases}$$

and M is a constant. Then (1) represents a linear transformation of X into itself.

Example 4. Let $A = (a_{nk})_{n,k=1}^\infty$ be an infinite matrix such that $a_{nk} \to 0$ ($n \to \infty$, k fixed) and $\sup_n \sum_k |a_{nk}| < \infty$. Then A defines a linear transformation on c_0 into itself, where $(A_n(x))$ is defined as $A_n(x) = \sum_{k=1}^\infty a_{nk} x_k$ provided the series converges for each n.

8.3 Isomorphism
Two mathematical structures are equivalent if they can be put into one-to-one correspondence in a way that preserves structure. For example, the concept of isometric metric spaces, i.e. the only difference between two isometric metric spaces is in the names given to the elements in their underlying sets.

An analogous situation arises when we deal with linear spaces. Let us make this concept precise.

Definition 4. The linear spaces X and Y over the same scalar field $\mathbf{K}$ are said to be *isomorphic* (or linearly isomorphic) if there exists bijective linear mapping T between X and Y. The mapping T is then said to be an *isomorphism*.

Obviously T puts X and Y into one-to-one correspondence and it preserves linear space structure.

Note that there is no demand in Definition 4 that the isomorphism between two isomorphic linear spaces be unique.

Example 5. Let $X = \mathbb{R}^2$ and let Y be the linear space made up of all functions f defined on $[0, T]$ of the form $f(t) = a_0 + a_1 t$, where a_0 and a_1 are arbitrary reals. The linear space structure on Y is of course, defined in the obvious way. These two linear spaces are isomorphic. One isomorphism, say T_1, mapping $\mathbb{R}^2$ onto Y is defined by $T_1(x) = x_1 + x_2 t$, where $x = (x_1, x_2)$.

Example 6. Let cs: $= \left\{ x = (x_k): \sum_k x_k \text{ converges} \right\}$, i.e. the space of convergent series. Then cs is linearly isomorphic to c with the usual linear operations in sequence spaces.

Define f:cs $\to c$ by $f(a) = A$, where $a = (a_k) \in$ cs and $A = (A_k) \in c$ with $A_k = a_1 + a_2 + \ldots + a_k$ the kth partial sum of $\sum_k a_k$.

The linearity of f is a consequence of the fact that $\sum_{k=1}^{n} (\lambda a_k + \mu b_k) = \lambda A_n + \mu B_n$ for every n. Now if $f(a) = f(b)$ then $A_k \doteq B_k$ for every k and thus $a_1 = b_1, a_2 = b_2, \ldots$, i.e. $a = b$. Hence f is injective.

If $A \in c$ is given, take $a_1 = A_1, a_2 = A_2 - A_1, \ldots, a_k = A_k - A_{k-1}$ for $k > 1$. Then $\sum_{i=1}^{k} a_i = A_k \to l (k \to \infty)$ and hence $\sum_i a_i$ converges; i.e. $a \in$ cs, clearly $f(a) = A$. Hence f is surjective. Therefore, f is isomorphism between cs and c.

8.4   Linear Dependence and Linear Hull

Definition 5. Let S be a subset of a linear space X. A point $x \in X$ is said to be a *linear combination* of points of S if there exists a finite set of points $\{x_1, x_2, \ldots, x_n\}$ in S and a finite set of scalars $\{\alpha_1, \alpha_2, \ldots, \alpha_n\}$ such that

$$x = \alpha_1 x_1 + \alpha_2 x_2 + \ldots + \alpha_n x_n \tag{1}$$

If the set S is empty, we agree that the origin θ is the unique point that is a linear combination of points in S. It should be noted that the expression for x in (1) may not be uniquely determined.

Example 7. Let $X = C[0, 1]$, and let S be the infinite set containing the continuous functions $\{1, t, t^2, \ldots\}$. The set of all linear combinations of points in S is the set of all polynomials in t, i.e. all functions of the form

$$x(t) = a_0 + a_1 t + \ldots + a_n t^n, t \in [0, T]$$

where $a_0, a_1, \ldots, a_n$ are scalars and $n = 0, 1, 2, \ldots$.

Definition 6. A set $\{x_1, x_2, \ldots, x_n\}$ in a linear space X is said to be *linearly independent* if a relation of the form $\lambda_1 x_1 + \lambda_2 x_2 + \ldots + \lambda_n x_n = 0$ implies that $\lambda_1 = \lambda_2 = \ldots = \lambda_n = 0$. If it is not linearly independent it is called *linearly dependent*.

An arbitrary subset 'not necessarily finite' of X is called linearly independent iff every one of its finite subsets is linearly independent.

Note that the empty set $\varnothing$ is linearly independent.

Example 8. Let $X = \mathbb{C}^n$ and $e_k = \{0, 0, \ldots, 0, 1, 0, 0, \ldots\}$ where

1 occurs at the kth position and zeros elsewhere. Then the set of unit vectors $\{e_1, e_2, \ldots, e_n\}$ is linearly independent, since

$$\lambda_1 e_1 + \lambda_2 e_2 + \ldots + \lambda_n e_n = \{\lambda_1, \lambda_2, \ldots, \lambda_n\}$$

and $\lambda_1 e_1 + \lambda_2 e_2 + \ldots + \lambda_n e_n = \theta \Rightarrow \lambda_1 = \lambda_2 = \ldots = \lambda_n = 0$

Definition 7. Let S be a subset of the linear space X. Then l.hull (S), the *linear hull* of S, is the intersection of all subspaces containing S.

We often refer l.hull (S) as 'span of S' or 'subspace generated by S'.

Example 9. In Example 8, we saw that the set $S = \{e_1, e_2, \ldots, e_n\}$ is linearly independent. Now if $x = (x_1, x_2, \ldots x_n) \in \mathbb{C}^n$, then $x = x_1 e_1 + x_2 e_2 + \ldots + x_n e_n$ is a linear combination of elements of S. Hence $x \in$ l.hull (S), so that $\mathbb{C}^n =$ l.hull (S).

Example 10. Let $X = \{0\}$ be a trivial linear space, then $\varnothing$ is linearly independent and l.hull $(\varnothing) = \{0\}$.

8.5 Hamel Basis and Dimensionality

A Hamel basis is the natural concept of basis for spaces that have linear structure only and it allows us to distinguish between finite and infinite dimensional linear spaces.

Definition 8. A set B in a linear space X is said to be a *Hamel basis* for X if

(i) B is linearly independent set and
(ii) l.hull $(B) = X$.

Definition 9. The cardinal number of any Hamel basis of a linear space X is said to be the *dimension* of X. We denote the dimension of X by dim (X).

Example 11. In the plane a set B containing any two noncollinear vectors is a Hamel basis of coordinate system for the plane.

Example 12. In Example 9, we have seen that $\mathbb{C}^n =$ l.hull (S). Since S is a finite set it follows that $\mathbb{C}^n$ is a finite dimensional. S is a Hamel base for $\mathbb{C}^n$, and dim $\mathbb{C}^n = n$.

Example 13. Let $X = l_2 := \left\{ x = (x_k) : \sum_k |x_k|^2 < \infty \right\}$ and let $A = \{e_1, e_2, \ldots, e_n\}$. It is easily seen that A is linearly independent. Hence, we may suspect that A is a basis in the sense of above Definition 8. *It is not!* Since we allow ourselves only finite sums, $a_1 x_1 + a_2 x_2 + \ldots + a_n x_n$, we see that l.hull $(A) \neq X$.

Now, we give two important results.

Theorem 1. If X is finite dimensional, with $\dim(X) = n$, then X is isomorphic to $\mathbb{C}^n$.

Proof. Since X is finite dimensional, there is a Hamel base $\{b_1, b_2, \ldots, b_n\}$. If $x \in X$ then $x = \lambda_1 b_1 + \lambda_2 b_2 + \ldots + \lambda_n b_n$ for some scalars λ_i. The λ_i are unique, for if $x = \mu_1 b_1 + \mu_2 b_2 + \ldots + \mu_n b_n$ then $(\lambda_1 - \mu_1)b_1 + \ldots + (\lambda_n - \mu_n)b_n = \theta$. Hence $\lambda_i = \mu_i$ $(1 \leq i \leq n)$, by linear independence of b_i. It follows now that the map $f{:}X \to \mathbb{C}^n$, given by $f(x) = (\lambda_1, \lambda_2, \ldots, \lambda_n)$ is well-defined. It is easy to check that f is bijective and $f(\alpha x + \beta y) = \alpha f(x) + \beta f(y)$ for scalars α, β and $x, y \in X$. Hence, f is an isomorphism.

Theorem 2. Every linear space X has a Hamel base.

Proof. Let P be the class of all linearly independent subsets of X. Then P is nonempty since $\varnothing \in P$. Partially order P by set inclusion and let $T = \{L_\alpha\}$ be a totally ordered subset of P. Let $L = \cup L_\alpha$. Then L is a linearly independent set. For if $S = \{s_1, \ldots, s_n\}$ is a finite subset of L then $s_i \in L_{\alpha_i}$ $(1 \leq i \leq n)$. By the total order in T we may arrange that $L_{\alpha_1} \subset \ldots \subset L_{\alpha_n}$. Hence $s_i \in L_{\alpha_n}$ $(1 \leq i \leq n)$, so that S is linearly independent, being a finite subset of the linearly independent L_{α_n}. Thus T has L for an upper bound. By Zorn's lemma, P has a maximal element, say B. Now every x in X is also in l.hull (B). For if not then there exists an $x \in$ l.hull $(B)^c$, and hence $B \cup \{x\}$ is linearly independent. But $B \cup \{x\} \supset B$, contrary to the fact that B is maximal. Therefore, B is linearly independent and l.hull $(B) = X$. Hence B is the required Hamel base.

8.6   Convexity and Other Related Concepts

We give some important properties of sets in a linear space.

Definition 10. Let E be a subset of a linear space. Then

(i) E is called *convex* if $x, y \in E$, $\lambda + \mu = 1$, $\lambda \geq 0$, $\mu \geq 0$ imply $\lambda x + \mu y \in E$, i.e. if $\lambda E + \mu E \subset E$. In other words, a set is convex iff it includes the line segment joining any two of its points.

(ii) E is called *absorbing* if for each vector x, there exists $\varepsilon > 0$ such that $\lambda x \in E$, wherever λ is scalar satisfying $|\lambda| \leq \varepsilon$. We say that E is *absorbing at* a if $E - a$ is absorbing.

(iii) E is called *balanced* if $x \in E$, $|\lambda| \leq 1$ imply $\lambda x \in E$, i.e. $\lambda E \subset E$.

(iv) E is called *absolutely convex* if $x, y \in E$, $|\lambda| + |\mu| \leq 1$ imply $\lambda x + \mu y \in E$.

Note that every subspace is absolutely convex and every absolutely convex set is convex.

(v) Convex, balanced and absorbing set is called a *balloon*.

Example 14. (i) Empty set is balanced and convex.

(ii) The following set is sometimes called *Schatz's apple*. It is the plane set A consisting ι^f the union of two closed unit disks centred at $(1, 0)$, $(-1, 0)$, and the line segment joining $(0, -1)$ to $(0, 1)$. Let us consider $\mathbb{R}^2$. Then A is absorbing since every line through the origin meets it in an interval of positive length. Next, consider the plane as $\mathbb{C}$, a (complex) linear space. Now A is no longer absorbing, since A includes no disk of positive radius centred at the origin.

(iii) Let A be a nonsquare, plan rectangle (with interior) centred at 0. Then A is balanced as a subset of the real linear space $\mathbb{R}^2$, but A is not balanced as a subset of the complex linear space $\mathbb{C}$, since for example $iA \not\subset A$.

(iv) Let d be the usual metric on $\mathbb{C}^n$. Then every sphere $S_r(a)$ in $\mathbb{C}^n$ is convex. For if $d(x, a) < r$, $d(y, a) < r$, $\lambda + \mu = 1$, then by Minkowski inequality

$$d(\lambda x + \mu y, a) = \left(\sum_{k=1}^{n} |\lambda(x_k - a_k) + \mu(y_k - a_k)|^2 \right)^{1/2}$$

$$\leq \lambda d(x, a) + \mu d(y, a) < r$$

(v) Sphere centred at 0 in $\mathbb{R}^2$, $\mathbb{R}^3$, $\mathbb{C}$ are balloons.

Definition 11. A nonempty set A is called *affine* if $\lambda A + (1 - \lambda) A \subset A$ for all scalars λ.

Thus an affine set is convex and a linear subspace is affine.

Definition 12. A *cone* is a nonempty set K which has the property that $\lambda K \subset K$ for every scalar $\lambda \geq 0$.

Example 15. Any linear subspace is a cone. An example in $\mathbb{C}$ is the region between two rays issuing from the origin, including neither, one, nor both rays; a special case of this is the upper half plane, including the positive real axis.

Note that a cone K is convex iff $K + K \subset K$.

8.7 Seminorms

There is a simple way of generating absolutely convex sets using certain types of real functions called seminorms.

Definition 13. A *seminorm* p, on a linear space X, is a function $p: X \to \mathbb{R}$ such that

(i) $p(\lambda x) = |\lambda|\, p(x)$, for all $\lambda \in \mathbb{R}$ and all $x \in X$ (absolute homogeneity),

(ii) $p(x + y) \le p(x) + p(y)$ for all $x, y \in X$ (subadditivity).

Note that p is always non-negative : By (i) and (ii) we have

$$0 = p(\theta) \le (x) + p(-x) = 2p(x). \text{ Hence } p(x) \ge 0.$$

Example 16. (i) $p(x) = |x|$ is a seminorm on $\mathbb{C}$.

(ii) $p_1(x) = \sup_n |x_n|$ and $p_2(x) = \left|\lim_n x_n\right|$ both are seminorms on c.

(iii) If $f: X \to \mathbb{C}$ is a linear map, then $p(x) = |f(x)|$ is a seminorm on X. Using linearity of f, we have

$$p(\lambda x) = |f(\lambda x)| = |\lambda f(x)| = |\lambda||f(x)| = |\lambda|\, p(x)$$

and $p(x + y) = |f(x + y)| = |f(x) + f(y)| \le |f(x)| + |f(y)| = p(x) + p(y)$.

(iv) $p(a) = \sum_n |a_n|$ is a seminorm on l_1.

Theorem 3. Let p be a seminorm on a linear space X and let $r > 0$. Then the sets $\{x : p(x) < r\}$ and $\{x : p(x) \le r\}$ are absolutely convex.

Proof. Suppose that $p(x) \le r$, $p(y) \le r$. Then

$$p(\lambda x + \mu y) \le |\lambda|\, p(x) + |\mu|\, p(y) \quad (\because p \text{ is seminorm})$$

$$\le (|\lambda| + |\mu|)r$$

$$\le r, \text{ whenever } |\lambda| + |\mu| \le 1$$

Hence $\{x : p(x) \le r\}$ is absolutely convex. Similarly, we can prove for $\{x : p(x) < r\}$.

8.8   Linear Topological and Linear Metric Spaces

Definition 14. A *linear topological space* is a linear space X which has a topology T, such that the algebraic operations of addition and scalar multiplication in X are continuous.

Continuity of addition means that $f : X \times X \to X$ defined by $f(x + y) = x + y$ is continuous on $X \times X$, and by continuity of scalar multiplication we mean that $f : \mathbb{C} \times X \to X$ defined by $f(\lambda, x) = \lambda x$ is continuous on $\mathbb{C} \times X$.

Example 17. In $\mathbb{C}$, we choose any points a and b and then take x close to a (i.e. $d(x, a)$ small) any y close to b, we shall have $x + y$ close to $a + b$, since

$$d(x + y, a + b) = |x + y - (a + b)|$$

$$= |(x - a) + (y - b)| \leq d(x, a) + d(y, b)$$

This situation may be described by saying that the operation (+) of addition in $\mathbb{C}$ is continuous.

Likewise, if we choose any $\lambda_0 \in \mathbb{C}$ and any $a \in \mathbb{C}$ and then take λ close to λ_0 and any x close to a, we shall have λx close to $\lambda_0 a$, since

$$d(\lambda x, \lambda_0 a) = |\lambda x - \lambda_0 a|$$

$$= |(\lambda - \lambda_0)(x - a) + \lambda_0(x - a) + (\lambda - \lambda_0)a|$$

$$\leq |\lambda - \lambda_0| \, d(x, a) + |\lambda_0| d(x, a) + |\lambda - \lambda_0||a|$$

Hence, we may say that the operation (·) of scalar multiplication in $\mathbb{C}$ is continuous.

Definition 15. A metric d on a linear space X is *translation invariant* if

$$d(x + z, y + z) = d(x, y) \text{ for all } x, y, z \in X$$

Example 18. (i) $d(x, y) = |x^3 - y^3|$ defines a metric on the real linear space $\mathbb{R}$, but d is not translation invariant.

For example, let $x = 1$, $y = 0$, $z = 1$.

Then $\qquad d(1 + 1, 0 + 1) - d(1, 0) = d(2, 1) - d(1, 0)$

$$= 7 - 1 = 6 \neq 0$$

Hence $\quad d(1 + 1, 0 + 1) \neq d(1, 0)$.

(ii) $d(x, y) = \sup_k |x_k - y_k|^{p_k}$ defines a metric on $l_\infty(p)$, where $p = (p_k) = (1/k)$. Here d is also translation invariant, since

$$d(x + z, y + z) = \sup_k |(x_k + z_k) - (y_k + z_k)|^{p_k}$$

$$= \sup_k |x_k - y_k|^{p_k} = d(x, y)$$

Definition 16. A *linear metric space* (X, d) is a linear space X with a translation invariant metric d on X such that addition and scalar multiplication are continuous in (X, d).

Example 19. The space $l(p)$ with the metric $d(x, y) = \sum_k |x_k - y_k|^{p_k}$ is a linear metric space, where $0 < p_k \leq 1$ for all $k \in \mathbb{N}$. It is easily checked that d is a metric on $l(p)$. Also $l(p)$ is a linear subspace of w, since ·

$$|x_k + y_k|^{p_k} \le |x_k|^{p_k} + |y_k|^{p_k}$$

and $$|\lambda x_k|^{p_k} \le \max\{1, |\lambda|\} \cdot |x_k|^{p_k}$$

Now, for $x, y, z \in l(p)$

$$d(x + z, y + z) = \sum_k |(x_k + z_k) - (y_k + z_k)|^{p_k}$$

$$= \sum_k |x_k - y_k| = d(x, y)$$

i.e. d is translation invariant.

Addition is continuous, since for any $a, b \in l(p)$, if $x, y \in l(p)$ then

$$d(x + y, a + b) = \sum_k |(x_k - y_k) - (a_k + b_k)|^{p_k}$$

$$\le \sum_k |x_k - a_k|^{p_k} + \sum_k |y_k - b_k|^{p_k}$$

$$= d(x, a) + d(y, b)$$

Hence $d(x + y, a + b) < \varepsilon$ if $d(x, a) + d(y, b) < \delta(= \varepsilon)$. Finally, we show that scalar multiplication is continuous.

$$d(\lambda x, \lambda_0 a) = \sum_k |\lambda x - \lambda_0 a|^{p_k}$$

$$= \sum_k |(\lambda - \lambda_0)(x_k - a_k) + \lambda_0(x_k - a_k) + (\lambda - \lambda_0)a_k|^{p_k}$$

$$\le \sum_k |\lambda - \lambda_0|^{p_k}|x_k - a_k|^{p_k} + \sum_k |\lambda_0|^{p_k}|x_k - a_k|^{p_k}$$

$$+ \sum_k |\lambda - \lambda_0|^{p_k}|a_k|^{p_k}$$

$$= \Sigma' + \Sigma'' + \Sigma''', \text{ say}$$

Now $|\lambda_0|^{p_k} \le \max\{1, |\lambda_0|\} = M$, say for all $k \in \mathbb{N}$. First we choose λ such that $|\lambda - \lambda_0| < 1$, so that we then have $|\lambda - \lambda_0|^{p_k} < 1$. Then $\Sigma' \le d(x, a)$ and $\Sigma'' \le M d(x, a)$.

Now, let $m \in \mathbb{N}$. Then $\Sigma'' \le \sum_{k=1}^{m} |\lambda - \lambda_0|^{p_k}|a_k|^{p_k} + \sum_{k=m+1}^{m} |a_k|^{p_k} = A + B$. Take any $\varepsilon > 0$. Then since $\sum_k |a_k|^{p_k}$ converges we may choose $m = m(\varepsilon, a) \in \mathbb{N}$ such that $B < \varepsilon/3$. Having chosen m we then have $A \to 0$ as $\lambda \to \lambda_0$. Hence $A < \varepsilon/3$ if $|\lambda - \lambda_0|$ is small enough, say $|\lambda - \lambda_0| < \alpha$ for some α, with $0 < \alpha < 1$. Consequently, defining

$$\delta = \min\left\{\alpha, \frac{\varepsilon}{3}(1 + M)\right\}, \text{ if } |\lambda - \lambda_0| + d(x, a) < \delta, \text{ then}$$

$$d(\lambda x, \lambda_0 a) \le d(x, a) + Md(x, a) + \frac{\varepsilon}{3} + \frac{\varepsilon}{3}$$

$$= (1 + M)\, d(x, a) + \frac{2}{3}\varepsilon < \frac{\varepsilon}{3} + \frac{2\varepsilon}{3} = \varepsilon$$

i.e. scalar multiplication is continuous. Hence $l(p)$ is a linear metric space.

8.9 Paranormed Spaces

The concept of paranorm is a generalization of that of absolute value. The paranorm of x may be thought of as the distance from x to θ.

Definition 17. A *paranorm* is a function $g: X \to \mathbb{R}$ defined on a linear space X such that for all $x, y, z \in X$

(P1) $g(x) = 0$ if $x = \theta$

(P2) $g(-x) = g(x)$

(P3) $g(x + y) \le g(x) + g(y)$

(P) If $\{\lambda_n\}$ is a sequence of scalars with $\lambda_n \to \lambda_0$ $(n \to \infty)$ and x_n, $a \in X$ with $x_n \to a$ $(n \to \infty)$, in the sense that $g(x_n - a) \to 0$ $(n \to \infty)$, then $\lambda_n x_n \to \lambda_0 a$ $(n \to \infty)$, in the sense that $g(\lambda_n x_n - \lambda_0 a) \to 0$ $(n \to \infty)$.

A paranorm g for which $g(x) = 0$ implies $x = \theta$ is called a *total paranorm* on X, and the pair (X, g) is called a *totally paranomed space*.

If we suppose that (X, d) is a linear metric space and for each $x \in X$ let us define $g(x) = d(x, \theta)$ where θ is the zero element in X, then it is straightforward to check that the properties of d imply the above properties of g.

Remark (i) If (X, g) is a totally paranormed space, then it readily follows that d defined by $d(x, y) = g(x - y)$ is such that (X, d) is a linear metric space.

(ii) A linear metric space and a totally paranormed space are really the same thing, like wise linear semimetric space and a paranormed space are equivalent.

Theorem 4. Each seminorm p on X is a paranorm, but converse need not be true.

Proof. From condition (i) of the definition of seminorm, we have

$$p(\theta) = p(\theta \cdot x) = |\theta|\, p(x) = 0 \text{ and } p(-x) = |-1|p(x) = p(x)$$

Also (P3) is same as (ii) of the definition of seminorm. Now

$p(\lambda_n x_n - \lambda_0 a)$

$$= p(\lambda_n x_n + \lambda_0 x_n - \lambda_0 x_n + \lambda_n a - \lambda_n a - \lambda_0 a + \lambda_0 a - \lambda_0 a)$$

$$= p([\lambda_n x_n - \lambda_n a - \lambda_0 x_n + \lambda_0 a] + [\lambda_0 x_n - \lambda_0 a] + [\lambda_n a - \lambda_0 a])$$

$$\leq p((\lambda_n - \lambda_0)(x_n - a)) + p(\lambda_0(x_n - a)) + p((\lambda_n - \lambda_n - \lambda_0)a)$$

$$= |\lambda_n - \lambda_0| p(x_n - a) + |\lambda_0| p(x_n - a) + |\lambda_n - \lambda_0| p(a)$$

Thus, as $n \to \infty$, $\lambda_n \to \lambda_0$ and $p(x_n - a) \to 0$ imply $p(\lambda_n x_n - \lambda_0 a) \to 0$, i.e. p is a paranorm.

Conversely, if we consider $g(x) = \sum_k |x_k|^{p_k}$ on $l(1/k)$, then g is a paranorm. Clearly, there is $x = \{0,1, 0, 0, \ldots\}$ such that $g(2x) < 2g(x)$. Therefore, g is not absolutely homogeneous. Hence, g is not a seminorm.

Example 20. $l(p)$ is a totally paranormed for any $p = \{p_k\} \in l_\infty$. It is easily checked that $l(p)$ is a linear space.

To get paranorm, define $g : l(p) \to \mathbb{R}$ by $d(x, y) = g(x - y)$, where $x, y \in l(p)$, and d is a metric on $l(p)$ given by

$$d(x, y) = \left(\sum_k |x_k - y_k|^{p_k}\right)^{1/M}$$

where $M = \max\{1, H\}$, $0 \leq p_k < \sup p_k = H < \infty$.

Then $$g(x) = d(x, \theta) = \left(\sum_k |x_k|^{p_k}\right)^{1/M}$$

and $g(-x) = \left(\sum_k |-x_k|^{p_k}\right)^{1/M} = \left(\sum_k |-1|^{p_k} |x_k|^{p_k}\right)^{1/M} = g(x)$

$$g(\theta) = 0, \text{ since } g(\theta) = d(\theta, \theta) = 0$$

and $$g(x) = 0 \Rightarrow d(x, \theta) = 0 \Rightarrow x = \theta$$

Now, by Minkowski's inequality

$$g(x + y) = \left(\sum_k |x_k + y_k|^{p_k}\right)^{1/M} \leq \left(\sum_k |x_k|^{p_k}\right)^{1/M} + \left(\sum_k |y_k|^{p_k}\right)^{1/M}$$

i.e. $g(x + y) \leq g(x) + g(y)$.

Condition (P4) can be shown easily by using Example 19. Hence $l(P)$ is a totally paranormed space.

8.10 FK and Other Spaces

One of the main features of the theory of FK-spaces is that it provides easy and short proofs of many classical results of summability theory.

Definition 18. Let (X, T) be a linear topological space and $x \in X$. Then, a set U is called a *neighbourhood* of x if there is an open set G with $x \in G \subset U$. Thus any open set G containing x is a neighbourhood of x.

A linear topological space is called *locally convex* iff every U contains an absolutely convex set V.

Note that a seminormed space is locally convex.

Definition 19. (i) A sequence space X with linear topology is called a *K-space* if each of the maps $P_i : X \to \mathbb{C}$ defined by $P_i(x) = x_i$ is continuous for $i = 1, 2, \ldots$

(ii) A *Fréchet space* is a complete linear metric space, or equivalently, a complete totally paranormed space.

In other words, a locally convex space is called a Fréchet space if it is metrizable and the underlying metric space is complete.

Note that w is a Fréchet space with its usual metric.

(iii) K-space X is called an *FK-space* if X is a complete linear metric space. In other words, we say that X is an FK-space if X is a Fréchet space with continuous coordinate projection, we mean if $x^{(n)} \to x(n \to \infty)$ in the metric of X then $x_k^{(n)} \to x_k(n \to \infty)$ for each $k \in \mathbb{N}$, i.e. for each $k \in \mathbb{N}$, the linear functional $P_k(x) = x_k$ is such that P_k is continous on X, i.e. X is K-space.

Note that w is a locally convex FK-space with its usual metric.

Definition 20. Let H be a Hausdorff space and X a linear space. An *FH-space* is a Fréchet space X such that

(i) X is a linear subspace of H

(ii) The topology of X is stronger than that of H

Note that the letters F and H are in honor of M. Fréchet and F. Hausdorff.

Remark. FK-space is a special kind of FH-space in which $H = w$.

Examples 21. (i) The sequence spaces l_p $(p > 0)$, c_0, c and l_∞ are FH-spaces.

(ii) Let $A = (a_{nk})_{n,k=1}^{\infty}$ be a triangular matrix, then c_A is an FK-space, where

$$c_A := \{x = \{x_k\} : Ax \in c\}$$

Now, we prove an important result:

Theorem 5. An FK-space X contains l_1 iff

$$\{e^k : k = 0, 1, 2, \ldots\} \tag{1}$$

is a bounded subset of X.

Proof. Let X contain l_1. Then the inclusion map $l_1 \to X$ is continuous. Since $\{e^k : k = 0, 1, 2, \ldots\}$ is bounded in l_1, it is bounded in X.

Conversely, suppose (1) holds. Let $x = \{x_k\} \in l_1$ and let P be a continuous seminorm on X. Then

$$P\left(\sum_{k=m}^{n} x_k e^k\right) \leq \sum_{k=m}^{n} |x_k| P(e^k)$$

Since $x \in l_1$, i.e. $\sum_{k=0}^{\infty} |x_k|$ is convergent, we have

$$\sum_{k=m}^{n} |x_k| P(e^k) \to 0 \; (m, n \to \infty)$$

Thus $\left\{\sum_{k=0}^{n-1} x_k e^k\right\}$ is a Cauchy sequence convergent to x in X.

Moreover $x \in X$ since X is complete. Hence $l_1 \subset X$.
We derive the following corollary:

Corollary 6. Let $A = (a_{nk})_{n,k=1}^{\infty}$ be an infinite matrix then $A: l_1 \to X$ (X an FK-space) if and only if the columns of A belong to X and form a bounded subset of X.

We will demonstrate how this result can be used to get an easy and short proof of the following classical results of summability theory.

Theorem 7. $A:l_1 \to c$ iff

$$\sup_{n,k} |a_{nk}| < \infty \tag{1}$$

$$\lim_{n} a_{nk} \text{ exists for all } k \tag{2}$$

Putting $X = c$ in the above corollary, we get the result directly.

Theorem 8. $A : l_1 \to l_1$ iff $\sup_{k} \sum_{n} |a_{nk}| < \infty$
This can be obtained directly by taking $X = l_1$.

Similarly, if we take $X = l_p$ we get

Theorem 9. $A: l_1 \to l_p$ iff

$$\sup_k \sum_n |a_{nk}|^p < \infty, \text{ for the case } 1 \le p < \infty \tag{1}$$

$$\sup_{n,k} |a_{nk}| < \infty \text{ for the case } p = \infty \tag{2}$$

8.11 Schauder Basis and AK-Property

A Hamel base is free of topology. In certain problems it is useful to have a concept of basis which allows us to express, uniquely, an element x as an infinite series

$$x = \sum_k \lambda_k b_k$$

This idea automatically involves convergence and have some kind of topology.

Definition 21. A *Schauder basis* for a linear metric space is a sequence $\{b_k\}$ such that for any vector x there exists a unique sequence $\{\lambda_k\}$ of scalars such that $x = \sum_{k=1}^{\infty} \lambda_k b_k$. The series $\sum_k \lambda_k b_k$ which converges to x is called the expansion of x.

We shall use the word 'basis' to mean 'Schauder basis'. For a finite-dimensional space the concepts of Schauder and Hamel bases coincide.

Let ϕ be the set of finite sequences $x = \{x_k\}$, i.e. x is called finite if $x_k = 0$ for almost all k.

In ϕ, if we get it, say, the metric of c_0, $\{e_k\}$ is both a Schauder and Hamel basis, where $e_1 = \{1, 0, 0, \ldots\}$, $e_2 = \{0, 1, 0, 0, \ldots\}$ etc.

Example 22. In c_0, $\{e_k\}$ is a Schauder basis, (since each $x \in c_0$ is $\sum_k x_k e_k$, the expansion is unique, for $\sum_k b_k e_k$ diverges if $b \notin c_0$, and converges, if $b \in c_0$ to b, not x, if $\{b_k\} \ne \{x_k\}$) but not a Hamel basis, since its span is ϕ, a proper subset of c_0. On the other hand, any Hamel besis of c_0 is uncountable and hence is automatically not a Schauder basis.

Example 23. $l(p)$ and w have $\{e_k\}$ as basis, under their natural paranorms

$$g(x) = \left(\sum_k |x_k|^{p_k} \right)^{1/M} \text{ on } l(p)$$

$$g(x) = \sum_k 2^{-k} (|x_k|/(1 + |x_k|)) \text{ on } w$$

Let us consider $l(p)$, Take any $x = \{x_k\} \in l(p)$. Put

$$y_n = x - \{x_1, x_2, \ldots, x_n, 0, 0, \ldots\}$$

then $y_n = x - \sum_{k=1}^{n} x_k e_k = \{0, 0, \ldots, 0, x_{n+1}, x_{n+2}, \ldots\}$

Hence

$$[g(y_n)]^M = \sum_{k=n+1}^{\infty} |x_k|^{p_k} \to 0 \ (n \to \infty), \text{ i.e. } x = \sum_k x_k e_k$$

This representation for x is unique. For if there is another representation

$$x = \sum_k \lambda_k e_k \text{ then } g\left(\sum_{k=1}^{n} (\lambda_k - x_k) e_k \right) \to 0 \ (n \to \infty)$$

Therefore, $\sum_{k=1}^{n} |\lambda_k - x_k|^{p_k} \to 0$ which implies $\lambda_k = x_k$ for all k.

Definition 21. If $f : X \to Y$, the *graph* of f

$$\{(x, f(x)) : x \in X\}$$

is a subset of $X \times Y$. If it is closed, f is said to have *closed graph*.

We quote here the revised version of two famous theorems:

Theorem 10. (Closed Graph Theorem). If X, Y are Fréchet spaces and f is linear and has closed graph, then f is continuous.

Remark: Any continuous map to a Hausdorff space has closed graph.

Theorem 11. (Banach-Steinhaus Theorem). Let X be an FK-space and $\{f_n\}$ a sequence of continuous functions $f_n : X \to \mathbb{C} \ (n = 1, 2, \ldots)$. If $f(x) = \lim_n f_n(x)$ for all $x \in X$, then $f : X \to \mathbb{C}$ is continuous.

Theorem 12. Let X be a Fréchet space, Y an FH-space and $f : X \to Y$ linear. If $f : X \to H$ (Hausdorff space) is continuous, then $f : X \to Y$ is continuous.

Proof. It is sufficient by Theorem 10 to show that f has closed graph. Let T_H be the topology of H restricted to Y. Then $f : X \to (Y, T_H)$ has closed graph by Theorem 10 (Remark). Now the graph of f is closed in $(X, T_X) \times (Y, T_H)$, hence in $(X, T_X) \times (Y, T_Y)$ since $T_Y \supset T_H$.

Corollary 13. Let X be a Fréchet space, Y an FK-space, $f : X \to Y$ a linear map and P_n the nth coordinate, i.e. $P_n(y) = y_n (y \in Y)$ for all $n = 0, 1, \ldots$ If each map $P_n \circ f : X \to \mathbb{C}$ is continuous, so is $f : X \to Y$.

Proof. Since $P_n \circ f : X \to \mathbb{C}$ is continuous for each n, the map $f : X \to w$ is continuous by the equivelence of the coordinatewise convergence and convergence in w. By Theorem 12, $f : X \to Y$ is continuous.

Theorem 14. Let $X \supset \phi$ be an FK-space and $a \in w$. If the series $\sum\limits_{k} a_k x_k$ converges for each $x \in X$, then the linear functional $f_a : X \to \mathbb{C}$ defined by $f_a(x) = \sum\limits_{k} a_k x_k$ for all $x \in X$ is continuous.

Proof. For each $n \in \mathbb{N}$, we define the linear functional $f_{a,n} : X \to \mathbb{C}$ by $f_{a,n}(x) = \sum\limits_{k=1}^{n} a_k x_k$ for all $x \in X$. Since X is an FK-space, the coordinates $P_k : X \to \mathbb{C}$ are continuous on X for all $k = 0, 1, \ldots$, and so are the functionals $f_{a,n} = \sum\limits_{k=1}^{n} a_k P_k$ $(n = 1, 2, \ldots)$ for each $x \in X$, $f_a(x) = \lim\limits_{n \to \infty} f_{a,n}(x)$ exists, and so $f_a : X \to \mathbb{C}$ is continuous by Theorem 11.

Theorem 15. Any matrix map between FK-spaces is continuous.

Proof. Let X and Y be FK-spaces, $A \in (X, Y)$ and the map $f_a : X \to Y$ be defined by $f_A(x) = A(x) \ \forall \ x \in X$. Since the maps $P_n \circ f_A : X \to \mathbb{C}$ are continuous for all $n \in \mathbb{N}$ by Theorem 14, the linear map f_A is continuous by Corollary 13.

Definition 22. An FK-space $X \supset \phi$ has AK if, for every sequence $x = \{x_k\} \in X$, $x = \sum\limits_{k=1}^{\infty} x_k e^{(k)}$, i.e. $x^{[n]} = \sum\limits_{k=1}^{n} x_k e^{(k)} \to x$ $(n \to \infty)$, where $x^{[n]}$ is the nth section of x, and X has AD if ϕ is dense in X. An FK-space $X \supset \phi$ is said to have AB if $\{x^{[n]}\}$ is a bounded set in X for every $x \in X$.

The notation AK arises from the German words Abschnitts-Konvergenz (Sectional Convergence), AD from Abschnitts-dicht (Section Dense), and AB from Abschnitts-beschrankte (Sectional Boundedness).

Remark. Every AK space has AD. The converse is not true in general.

Example 24. The spaces w, c_0 and $l_p (1 \leq p < \infty)$ have AK, but the spaces c and l_∞ do not.

Let us consider the case w. Let, for each $n = 0, 1, \ldots$, $e^{(n)}$ be the sequence with $e_n^{(n)} = 1$ and $e_k^{(n)} = 0$ for $k \neq n$. Then $\{e^{(n)}\}_{n=0}^{\infty}$ is a Schauder basis of w. More precisely, every sequence $x = \{x_k\}_{k=0}^{\infty} \in w$ has a unique representation

$$x = \sum\limits_{k=0}^{\infty} x_k e^{(k)}, \text{ i.e. } \lim\limits_{m \to \infty} x^{[m]} = x$$

Therefore w has AK.

EXERCISES

1. If X_1 and X_2 are linear spaces over the same scalar field, prove that X_1 and X_2 are isomerphic iff diam $(X_1) =$ diam (X_2).

2. Consider the following differential equation on $C^2[0, \infty)$, the set of continuous functions (real or complex valued) on $[0, \infty)$ with continuous second derivatives:

$$\frac{d^2x}{dt^2} + b\frac{dx}{dt} + cx = 0 \qquad (1)$$

   If X denotes the set of all solutions of (1) show that X is a linear subspace of $C^2[0, \infty)$ and that $\text{diam}(X) = 2$.

3. Let X be a finite dimensional space with $\dim(X) = n$. Show that every set containing $n + 1$ points is linearly dependent.

4. Show that if A is a set in a linear space X with l.hull $(A) = X$, then A contains a Hamel basis of X.

5. Show that c_0 is infinite dimensional.

6. Prove that a set is absolutely convex iff it is convex and balanced.

7. Show that $\left\{x \in w: |x_n| < \dfrac{1}{n} \text{ for all } n\right\}$ is not absorbing in c_0.

8. Let A be a non-empty open absolutely convex set in a linear topological space X. Then prove that $p(x) = \inf \{\lambda > 0: x \in \lambda A\}$ is a seminorm on X.

9. Show that for $l_\infty(p)$ with $d(x, y) = \sup_k |x_k - y_k|^{p_k}$, d is translation invariant and addition is continuous but scalar multiplication is not continuous, where $p = (p_k) = (1/k)$.

10. For $a \in l^p (0 < p < 1)$, let $g(a) = \sum_{k=1}^{\infty} |a_k|^p$. Show that g is a paranorm.

11. Let g be a paranorm on a linear space. Show that if $g(a - b) = 0$ then $g(a) = g(b)$ but not conversely.

12. If g is a paranorm such that $g(\lambda x) \leq |\lambda| g(x)$ for all scalars λ and all vectors x, then show that g is a seminorm.

13. Show that the map g defined by

$$g((x, y)) = |x|^p + |y|^p, \ 0 < p < 1$$

   defines a paranorm on $\mathbb{R}^2$.

14. In a linear metric space if the series $\sum_k x_k$ is convergent, then show that the sequence $\{x_k\}$ is convergent to zero, but not conversely.

15. Prove that the sequence space w is a Fréchet space with the paranorm defined by

$$g(x) = \sum_k 2^{-k} (|x_k|/(1 + |x_k|)), x = (x_k) \in w$$

16. Let $\{g_n\}$ be a sequence of paranorms on a linear space X. Show that

$$g(x) = \sum_k 2^{-k} g_k(x)/(1 + g_k(x))$$

is a paranorm on X. Prove that $g(x_n) \to 0$ $(n \to \infty)$ iff

$$g_k(x_n) \to 0 \ (n \to \infty) \text{ for each } k$$

17. Show that $c_0(p)$ is a linear metric space paranormed by

$$g(x) = \sup_k |x_k|^{p_k/M}, \ x = \{x_k\} \in c_0(p)$$

18. Show that $l_\infty(p)$ and $c(p)$ are not linear metric spaces. However, they turn out to be linear metric spaces iff inf $p_k > 0$.
19. Let (X, g) be a paranormed space with basis $B = \{b_k\}$. Prove that X is separable.
20. Prove that if an FK-space X has A K, then $z^{-1}X : = \{x \in w: x \cdot z \in X\}$ has also A K.

References

1. Copson, E.T. Metric spaces, Cambridge Univ. Press (1968).
2. Daneš, J. On the Istratesku's measure of noncompactness, Bull. Math. Soc. Sci. Math. R.S. Roumanie (N. S), 16 (1972), 403–406.
3. Darbo, G. Punti uniti in transformazioni a condominio non compactto, Rend. Sem. Univ. Padova, 24 (1955), 84–92.
4. Istrătesku, V. On a measure of noncompactness, Bull. Math. Soc. Sci. Math. R.S. Roumanie (N.S.), 16 (1972), 195–197.
5. Jain, P.K. and Ahmad, K. Metric spaces, Narosa Publishing House, New Delhi (1993).
6. Kuratowski, K. Sur les espaces complets, Fund. Math. 15 (1930), 301–309.
7. Maddox, I.J. Elements of Functional Analysis, Oxford Univ. Press (1970).
8. Malkowsky, E. and Rako č evic', V. An Introduction into the theory of sequence spaces and measures of noncompactness, 'Four Topics in Mathematics', Zborn-ik radova, 9 (17) (2000), 143–234.
9. Pitts, C.G.S. Introduction to Metric spaces, Oliver and Boyd (1972).
10. Reisel, R.B. Elementary Theory of Metric Spaces, Springer-Verlag, Berlin (1982).
11. Somasundram, D. and Choudhary, B. A First Course in Mathematical Analysis, Narosa Publishing House, New Delhi (1996).

Index